Measurement-based analysis of harmonic instabilities of single-phase photovoltaic inverters in public low voltage networks

Elias Kaufhold

Technische Universität Dresden

Measurement-based analysis of harmonic instabilities of single-phase photovoltaic inverters in public low voltage networks

Dipl.-Ing.
Elias Kaufhold

der Fakultät Elektrotechnik und Informationstechnik

der Technischen Universität Dresden

zur Erlangung des akademischen Grades
Doktoringenieur
(Dr.-Ing.)

genehmigte Dissertation

Tag der Einreichung: 04.09.2024
Tag der Verteidigung: 29.01.2025

Vorsitzender:	Prof. Dr. rer. nat. habil. Hans Georg Krauthäuser (TU Dresden)
Gutachter:	Prof. Dr.-Ing. Peter Schegner (TU Dresden)
Gutachter:	Prof. Sjef Cobben (Eindhoven University of Technology)
Weiteres Mitglied:	Prof. Dr.-Ing. Steffen Großmann (TU Dresden)

Bibliografische Information der Deutschen Nationalbibliothek: Die Deutsche Nationalbibliothek verzeichnet diese Publikation in der Deutschen Nationalbibliografie; detaillierte bibliografische Daten sind im Internet über http://dnb.dnb.de abrufbar.

Publisher: BoD · Books on Demand GmbH,
Überseering 33,
22297 Hamburg,
bod@bod.de

Printer: Libri Plureos GmbH, Friedensallee 273, 22763 Hamburg

ISBN: 978-3-7693-9930-1

Declaration

I declare that the work in this dissertation titled „Measurement-based analysis of harmonic instabilities of single-phase photovoltaic inverters in public low voltage networks“ has been carried out by me. This work has been supervised by

Prof. Dr.-Ing. Peter Schegner, TU Dresden (TUD), Dresden, Germany

who is also the director of the Institute of Electrical Power Systems and High Voltage Engineering (IEEH) at which this work was developed. The results have been achieved at the Chair of Electrical Power Supply in the working group Power Quality where

Prof. Dr.-Ing. habil. Jan Meyer, TU Dresden (TUD), Dresden, Germany

is the current host who supervised the related publications from a technical perspective and was specifically involved in the internal review of the papers. Furthermore,

Prof. Ph.D. Xiongfei Wang, Aalborg University (AAU), Aalborg, Denmark

has hosted and co-supervised me at his chair for power electronics system integration and materials during my research stay at Aalborg University (AAU), Denmark, from August 2021 until November 2021.
As this work has been mainly developed within the priority program of the Deutsche Forschungsgemeinschaft (DFG) „DFG SPP 1984: Hybrid and multimodal energy systems: System theoretical methods for the transformation and operation on complex networks“ in the projects „Novel Methods and Models for the Analysis of Harmonic Instabilities in Distribution Grids with a High Penetration of Power Electronics“ and „New techniques for the assessment of harmonic stability in public low voltage networks with very high share of distributed power electronic devices“, the results have been discussed with

Prof. Dr.-Ing. Johanna M.A. Myrzik, Uni Bremen (UB), Bremen, Germany

who was professor and director of the institute of automation technology at the cooperating university (UB) during the development of this work.
Other than the mentioned supervisors have not been involved in the presented work of this thesis. All information derived from previous, external work has been acknowledged with the respective references. The references [1–23] originate from the author. Furthermore, the references [24–27] have been supported by the author.

Elias Kaufhold

Elias Kaufhold
Thun, Switzerland, 16th March 2025

Acknowledgments

This work would not have been possible without the guidance and the provision of the financial funding to employ me during the past years by Mr. Prof. Dr.-Ing. habil. Peter Schegner. I thank him for the possibility to work at the institute and for being my doctorial and general supervisor.

Furthermore, Mr. Prof. Dr.-Ing. habil. Jan Meyer supervised me technically and made my stay in Aalborg in Denmark possible. His personal guidance provided me a support that challenged my work and opened up the opportunity to enter the scientific world to get the contacts, exchanges and feedback that have been required to fullfill the demands to answer the relevant research questions of the presented work. I thank him for accepting my curious and direct discussion mentality and guiding me into a more settled scientific discussion atmosphere as well as for his critical questions regarding my work to always enhance the quality.

I also thank my colleagues Morteza Pourarab and Shrinath Kannan who had to face my emotional ups and downs and sourrounded me with a calm and productive working environment and positive discussions.

In addition, I had an extensive support from my colleague Robert Stiegler regarding field measurements of public network impedances and also from my colleague Sascha Müller who also worked on modeling of power electronic devices. More general, I am thankful for all the support from the colleagues in the working group and the institute.

The great time at Aalborg university became possible thanks to Mr. Prof. Ph.D. Xiongfei Wang and the colleagues from Aalborg university, mainly Mr. Ph.D. Hong Gong and Mr Ph.D. Naser Nourani Esfetanaj.

I thank my students Ms. Qingqing Dong (Student and Diploma thesis), Mr. Simon Grandl (Student thesis), Mr. Philipp Werner (Student and Diploma thesis), Mr. Felix Hornung (Student thesis) and Ms. Sofia Barriuso-Gutierrez (Bachelor thesis) who trusted me in guiding them through their studies. Also, I am thankful for Mr. Jonas Marko's work on related studies to this work with regard to frequency coupling components of passive rectifiers during his time as student assistant at the institute under my supervision.

And lastly, I thank my parents, my family, Orell, Nora and friends who have been a mental support over the years that it took to complete this work.

Kurzfassung

Der Wandel von zentraler zu dezentraler Erzeugung elektrischer Energie führt zu einer ansteigenden Anzahl installierter, regenerativer Energieerzeuger auf Basis leistungselektronischer Geräte in öffentlichen Niederspannungsnetzen.
Die Interaktion zwischen leistungselektronischen Geräten und dem Stromversorgungsnetz, in der vorliegenden Arbeit das öffentliche Niederspannungsnetz, kann den Betrieb leistungselektronischer Geräte und von Betriebsmitteln im Netz erschweren.
Hersteller kennen in der Regel die Netzcharakteristika am Standort, an dem ein Gerät in Zukunft betrieben werden soll, nicht. Im Niederspannungsnetz können die Hintergrundspannung und die Netzimpedanz stark zwischen verschiedenen Netzen und Tageszeiten variieren.
Im Kleinleistungsbereich bis zu einer Bemessungsleistung von wenigen Kilowatt sind die Anzahl an verfügbaren Geräten und die Zahl der Hersteller auf dem Markt groß.
Netzbetreiber können die Hintergrundspannung, beziehungsweise die Spannung am Netzanschlusspunkt, sowie die Netzimpedanz für konkrete Anschlusspunkte ermitteln. Allerdings sind ihnen die installierten Geräte und deren Topologien und Parameter in der Regel unbekannt und unzugänglich, da die Hersteller die Parameter als Herstellergeheimnis oftmals nicht angeben.
Um die Klimaziele zu erreichen, sind regenerative Energieerzeuger in den letzten Jahren zunehmend wichtiger geworden. Photovoltaiksysteme stellen dabei heutzutage eine der dominierenden Technologien dar. Photovoltaiksysteme im Kleinleistungsbereich werden an das Stromversorgungsnetz über einphasige Inverter angeschlossen. Der stabile Betrieb dieser Inverter ist daher entscheidend für die Zuverlässigkeit und Verfügbarkeit zukünftiger Stromversorgungsnetze. Eine technische Herausforderung in Bezug auf den stabilen Betrieb der Inverter stellen harmonische Instabilitäten dar. Für die Analyse harmonischer Instabilitäten von Invertern wird die Interaktion zwischen der Regelung und dem netzseitigen Filterkreis des Inverters mit der Netzimpedanz des Stromversorgungsnetzes im harmonischen Frequenzbereich, das heißt über 50 Hz bis 2 kHz, untersucht.
In der vorliegenden Arbeit wird der Stand der Technik erweitert, um eine messbasierte Analyse und Bewertung bezüglich der harmonischen Stabilität unbekannter, einphasiger Inverter vornehmen zu können. Die untersuchten einphasigen Inverter sind kommerziell verfügbar. Weiterhin wurden Simulationsmodelle generiert. Die messbasierte Identifikationsmethodik wurde für Inverter weiter entwickelt und validiert. Neben theoretischen Testfällen wurden Impedanzen von realen Niederspannungsnetzen verwendet, um die harmonische Stabilität der Inverter zu bewerten. Abschließend werden Designempfehlungen anhand der Erkenntnisse der vorliegenden Arbeit abgeleitet und Grenzen der angewendeten Methodik in Bezug auf die Bewertung des allgemein stabilen Betriebs der Inverter präsentiert.

Abstract

The transition of power systems from central energy generation to decentralized energy generation leads to a growth of renewable energy generators with power electronic devices in public low voltage networks.
In practice, the interaction of a power electronic device and the power grid, i.e. the low voltage network, challenge the device operation, the operation of other grid-connected devices and the network equipment.
For a manufacturer, the network characteristics where the device is installed, are typically unknown, e.g. especially for devices in the low power range, e.g. up to some kW, where the number of manufactured devices is large. In addition, the characteristics vary largely between different low voltage networks and even between measurement points within the same low voltage network. On the other hand, for the network operators, the installed devices are unknown, i.e. a black-box, since the manufacturers keep the detailed device designs including the device parameters a manufacturer's secret.
With the aim to pursue the climate goals, renewable energy generators become more important. One of the main renewable energy generators are low power photovoltaic systems. These photovoltaic systems are connected to the grid via single-phase inverters. The stable operation of the inverter is consequently relevant for the reliability of power systems. One of the phenomena that challenge the stable operation of inverters are harmonic instabilities. The harmonic stability analysis identifies an instable operation of an inverter based on the interaction of the inverter control as well as the AC-side filter circuit and the network impedance in the harmonic frequency range, i.e. above 50 Hz up to 2 kHz.
In this dissertation, the currently known theory is extended to enable measurement-based assessments of the harmonic stability of unknown single-phase inverters for photovoltaic applications. The studied single-phase inverters are commercially available while in addition simulation models are developed. Advancements of the measurement-based model identification and the harmonic stability analysis are presented and validated. Next to theoretic test cases, impedance characteristics of real low voltage networks are also included to assess the harmonic stability of the inverters. As a conclusive result, device design recommendations are derived from the findings of the assessment and limitations of the harmonic stability with regard to the overall stable operation of the inverter are presented.

List of acronyms

AC	alternating current
AI	anti-islanding
ANF	active notch filter
ANN	artificial neural network
BSS	battery storage system
CM	condition monitoring
CPU	central processing unit
DC	direct current
DFG	Deutsche Forschungsgemeinschaft
DFT	discrete Fourier transform
DSO	distribution system operator
DUT	device under test
EMC	electromagnetic compatibility
ePLL	enhanced PLL
ESR	equivalent series resistance
EV	electric vehicle
FA	fast approach
FCM	frequency coupling matrix
FIR	finite impulse response
FLL	frequency locked loop
FS	frequency sweep
GB	gigabyte
HAM	harmonically coupled admittance matrix
HF	high frequency
HSS	harmonic state space
HTF	harmonic transfer functions
HV	high voltage
HW	Hammerstein-Wiener
IGBT	insulated gate bipolar transistor
IPT	inverse Park transform
LF	low frequency
LMS	least mean square
LTI	linear time-invariant
LTP	linear time-periodic
LV	low voltage
MC	Monte Carlo
MFC	mirror frequency coupled
MFD	mirror frequency decoupled
MPP	maximum power point

MPPT	maximum power point tracker
NOT	bitwise complement
ODE	ordinary differential equation
OSG	orthogonal signal generator
PE	power electronic
PI	proportional integral
PLL	phase locked loop
PoC	point of connection
PoW	point on wave
PR	proportional resonant
PV	photovoltaic
PWM	pulse width modulation
RAM	random-access memory
REM	Euclidean division
RHP	right half plane
RMS	root mean square
RoC	rate of change
SC	Study Committee
SOGI	second order generalized integrator
SPD	surge protection device
SVR	series voltage regulator
T4D	T4-delay
TD	time delay
TDC	total distortion current
THD	total harmonic distortion
THDS	subgroup total harmonic distortion
TP	two-pole
VEN	variable electrical network
VSC	voltage source converter
ZOH	zero-order hold

List of symbols

$\boldsymbol{A}(t)$	real-value system parameter matrix
$\boldsymbol{B}(t)$	real-value system parameter matrix
$\boldsymbol{C}(t)$	real-value system parameter matrix
C	capacitance
C_{DC}	DC-link capacitance
C_{f}	capacitance in the T-branch of an LCL-filter
D	delay
$\mathrm{devabs}_{\varphi Y}$	maximum absolute deviation of admittance phase angles
$\mathrm{devabs}_{\lvert \underline{Y} \rvert}$	maximum absolute deviation of admittance magnitudes
$\mathrm{devrel}_{\lvert \underline{Y} \rvert}$	maximum relative deviation of admittance magnitudes
$\boldsymbol{D}(t)$	real-value system parameter matrix
D_{simp}	approximated delay
$e_{\mathrm{Cap}}(t)$	energy stored in DC-link capacitance
$e_{\mathrm{Cap\,const}}$	constant part of energy stored in DC-link capacitance
$e_{\mathrm{Cap\,rip}}(t)$	time-dependent part of energy stored in DC-link capacitance
ε	manufacturing tolerance
$\mathrm{err}_{\mathrm{Inv}\,\lvert \underline{Y} \rvert}$	deviation of the impedance magnitude between the frequency sweep and the fast approach
$\mathrm{err}_{\varphi\,\mathrm{Inv}\,Y}$	deviation of the phase angle of the impedance magnitude between the frequency sweep and the fast approach
f_1	power frequency, i.e. the power system frequency
f_{C}	critical frequency at which the inverter becomes instable
f_{car}	carrier frequency of pulse width modulation
fcn	function
$\hat{f}$	function of Hilbert isomorphism
f_I	current frequency
$f_{LCL\mathrm{res}}$	resonance frequency of an LCL-filter
$f_{\mathrm{ref}\,h}$	harmonic frequency of reference signal for voltage at the inverter bridge
f_U	voltage frequency
$\boldsymbol{f_U}$	vector of applied voltage frequencies
Γ	Gamma function
gci	grid compatibility index
h	harmonic order
H_k	harmonic transfer function at order k
$\boldsymbol{H}$	harmonic transfer matrix
$i_{\mathrm{PoC}}(t)$	current at point of connection
$i_{\mathrm{PV}}(t)$	current from the PV-modules into the DC-side inverter clamps
J	Bessel function
K	order of model truncation, e.g. studied current frequencies

L	inductance
L_{C}	critical iductance for which the inverter becomes instable
L_{f}	overall inductance of an AC-side filter, e.g. an L-filter
$L_{\mathrm{f\,d}}$	device-side filter inductance of an LCL-filter
$L_{\mathrm{f\,g}}$	grid-side filter inductance of an LCL-filter
L_{g}	inductance of a public network
m	incremental parameter
m_{a}	modulation index
M_{mag}	magnetic coupling between two air coils
n	incremental parameter
N_{FS}	overall number of measurements of a frequency sweep
$n_{h\,\mathrm{FS}}$	number of tested harmonics of a frequency sweep
$n_{p\,\mathrm{FS}}$	number of tested operating powers of a frequency sweep
$n_{\varphi\,\mathrm{FS}}$	number of tested phase angles of a frequency sweep
$n_{\hat{u}_{\mathrm{FS}}}$	number of tested voltage amplitudes of a frequency sweep
$p_{\mathrm{Cap}}(t)$	power of DC-link capacitance
$p_{\mathrm{DC}}(t)$	power at DC-link side of the inverter bridge
$\varphi_{\mathrm{g}\,Z}$	phase angle of impedance of low voltage (LV) network
$\varphi_{\mathrm{Inv}\,Y}$	phase angle of inverter admittance
$\varphi_{\mathrm{Inv}\,Y\,\mathrm{FA}}$	phase angle of inverter admittance parametrized by the fast approach
$\varphi_{\mathrm{Inv}\,Y\mathrm{FS}}$	phase angle of inverter admittance parametrized by the frequency sweep
$\varphi_{\mathrm{Inv}\,Z}$	phase angle of inverter impedance
φ_{PM}	phase margin
$\varphi_{\mathrm{PoC}\,f1}$	phase angle between voltage and current at power frequency at the point of connection
$\varphi_{\mathrm{PoC\,ref}\,U\,f1}$	phase angle detected by the phase locked loop (PLL)
$\varphi_{\mathrm{PoC}\,U}$	phase angle between voltage at power frequency and voltage at a frequency f at the point of connection
p_{loss}	losses inside the inverter
$p_{\mathrm{PoC}}(t)$	power at point of connection
$p_{\mathrm{PoC\,av}}$	average power at point of connection
p_{PV}	power of photovoltaic modules
q	modulation parameter
q_h	adapted modulation parameter
R	resistance
$R_{\mathrm{f}\,C}$	damping resistance in the T-branch of an LCL-filter
$R_{\mathrm{f\,d}}$	parasitic filter resistance on the device side of an LCL-filter
$R_{\mathrm{f\,g}}$	parasitic filter resistance on the grid side of an LCL-filter
$R_{\mathrm{f}\,L}$	overall resistance of an AC-side filter, e.g. an L-filter
R_{g}	resistance of an LV network
σ_f	frequency resolution
σ_t	time resolution
$\mathbf{sws}(t)$	vector of switching signal
T	period duration
t	time
t_{FS}	duration to measure one measurement point in a frequency sweep
T_{sw}	switching time

$u_{\mathrm{DC}}(t)$	DC-link voltage
$u_{\mathrm{DC\,av}}$	averaged DC-link voltage
$u_{\mathrm{DC\,rip}}(t)$	voltage distortion of the DC-link voltage
$u_{\mathrm{g}}(t)$	background voltage of LV network
$u_{\mathrm{Inv}}(t)$	voltage at the inverter bridge
$u_{\mathrm{Inv\,ref}}(t)$	reference signal for voltage at the inverter bridge
$u_{\mathrm{PoC}}(t)$	voltage at point of connection
$u_{\mathrm{PV}}(t)$	voltage at DC-side inverter clamps
$\underline{U}_{Z\,\mathrm{test}}$	voltage over the test impedance
ω_{ff}	feed-forward angular frequency
ω_{PoC}	angular frequency of voltage at point of connection
$\boldsymbol{x}(t)$	state vector
$\underline{Y}_{11}$	admittance parameter of a two-port
$\underline{Y}_{12}$	admittance parameter of a two-port
$\underline{Y}_{21}$	admittance parameter of a two-port
$\underline{Y}_{22}$	admittance parameter of a two-port
$\mathbf{Y_{Inv}}$	frequency coupling matrix
$\lvert\underline{Y}_{\mathrm{Inv\,FA}}\rvert$	magnitude of inverter admittance parametrized by the fast approach
$\lvert\underline{Y}_{\mathrm{Inv\,FS}}\rvert$	magnitude of inverter admittance parametrized by the frequency sweep
$\underline{Z}_{\mathrm{f\,d}}$	device-side filter impedance of an LCL-filter
$\underline{Z}_{\mathrm{f\,g}}$	grid-side filter impedance of an LCL-filter
$\underline{Z}_{\mathrm{f}\,C}$	filter impedance of T-branch of an LCL-filter
$\underline{Z}_{\mathrm{g}}$	impedance connected to the AC side of the inverter, e.g. the grid side
$\underline{Z}_{\mathrm{Inv}}$	inverter impedance
$\underline{Z}_{\mathrm{test}}$	overall test stand impedance

Nomenclature

Symbol	Definition
x	scalar, real value
$\underline{x}$	complex value
$\hat{x}$	amplitude value
$y(x)$	dependency of y on x
X	scalar, root mean square value, magnitude, maximum counter
$\|x\|$	absolute value
$\boldsymbol{x}$	column vector
$\boldsymbol{X}$	matrix
$*$	convolution operator

Remark: Based on this nomenclature, time domain values are always explicitly marked by their dependency on time t, e.g. $u(t)$.

Contents

1 Introduction

1.1 Motivation

The demand for energy efficiency and the pursuit to obtain the climate goals leads to a growth of renewable energy generators and consequently an increasing share of installed power electronic (PE) devices in public LV networks [28]. One of the main applications that lead the transition from central energy generation to decentralized energy generation are photovoltaic (PV) systems. In LV networks, the number of installed devices is typically high and the rated power low. The device that connects a PV system to a power grid is an inverter. For low power applications, e.g. below 4.6 kW, a single-phase connection is typically allowed by national regulations and laws while applications with higher power ratings must be connected three-phase due to the induced unsymmetry. Consequently, low power inverters are designed single-phase for reasons of cost-efficiency. In the following, these single-phase inverters are simply called inverters implying the single-phase connection if not annotated otherwise.

Inverters have non-ideal voltage-current characteristics. Even for a sinusoidal voltage at the point of connection (PoC), i.e. ideal operation condition, the current response of the inverter is non-sinusoidal, i.e. the current waveform is distorted. The increasing number of installed inverters in energy networks affects the overall network characteristics, e.g. voltage distortions and network impedance characteristics at individual PoCs in the network. The voltage distortion levels at the PoCs and the respective network impedance characteristics seen by a grid-connected device determine the interaction of the network and this device. Interaction issues of PV inverters, i.e. highly distorted currents exceeding the allowed emission limits, have been reported repetitively in the past, e.g. in [29], and caused inverters to shut down. With regard to classical stability analysis, this phenomenon is called a harmonic instability, e.g. [30].

The first harmonic instability that has been reported occured in the Swiss railway grid [31]. The same physical effect has also been reported for the interaction of a large number of PV inverters and the public LV network in the Netherlands [32] in relation to resonance phenomena. Another harmonic instability has been reported at a German offshore windpark [33] where the resulting damage caused the entire windpark to stand still for five months at costs of 1.8 Mio € per day. More recently, field measurements have also been

documented in Switzerland that demonstrate the existing possibility of shut downs of PV inverters in public LV networks [34]. Shutdowns of PV inverters have been identified in a research measurement campaign in Australia [35] that are possibly related to inrush currents of transformers in the harmonic frequency range.
In practice, one of the challenges is the diversity of devices. Typically, the network operator does not know details about the installed grid-connected devices, e.g. the device parameters. On the other hand, the manufacturer of a device does not know the network characteristics. Because of large differences in the network characteristics (see appendix C), the manufacturer faces the challenge to design the control parameters for all possible power grids with one set of parameters or rather one overall device design.
From an abstract point of view, the analysis of the harmonic stability is an interdisciplinary research topic that interfaces various research areas, namely power electronics, power electronic systems, power systems, control and automation theory, system theory, simulation modeling and programming, measurements and signal processing.

1.2 Objective and structure

The objective of the following is the harmonic stability analysis of commercially available single-phase inverters in public LV networks.

1.2.1 Research questions

To specify the main research objective of this work, the following research question can be formulated:
How can the harmonic stability of commercially available single-phase inverters for photovoltaic applications be assessed?
To answer this main research question, two more research questions have to be solved in advance:

1. What are the input-output small-signal dependencies of inverters?
2. What are suitable measurement methods to identify representations of commercially available inverter?

1.2.2 Structure

Chapter 2 summarizes the state of the art with respect to the general operating principle of the inverter, modeling approaches, measurement-based identification methods, the harmonic stability analysis and open research gaps. Chapter 3 analyzes the linearity and the small-signal characteristics of inverters, specifically by studying the voltage-current relations of the inverter at the PoC including a deeper insight into the convolution of the control output signal and the physical signals at the inverter bridge. An improved depiction of the small-signal characteristics is proposed by the power cuboid for individual inverters. Chapter 4 uses the findings of chapter 3 to improve the measurement-based identification of the operating point dependent small-signal models of the inverter. Chapter 5 includes the results of the previous chapters and presents the measurement-based black-box harmonic stability assessment. Finally, chapter 6 concludes the overall results in terms of a summary and future work.

2 State of the art

2.1 Operating principle

To explain the operating principle of PV inverters that will simply be called inverters in the following of this work, a general scheme of an exemplary PV system, i.e. based on a PV inverter and PV modules, and the LV network with the background voltage $u_{\mathrm{g}}(t)$ and the network impedance $\underline{Z}_{\mathrm{g}}$ is depicted in figure 2.1. The inverter is connected to the LV network at the PoC with the respective voltage $u_{\mathrm{PoC}}(t)$, the current $i_{\mathrm{PoC}}(t)$ and the power $p_{\mathrm{PoC}}(t)$.

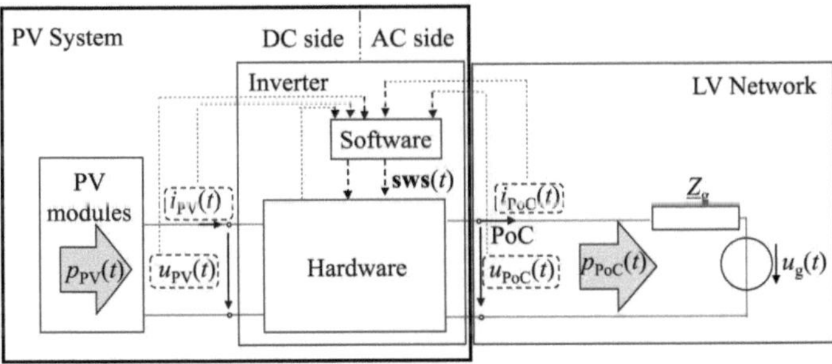

Figure 2.1: General scheme of a photovoltaic system and the low voltage network.

The inverter itself is structured into a direct current (DC) side and an alternating current (AC) side. The DC side is also called the application side or rather the inverter side and the AC side, e.g. the side where the inverter is connected to the LV network, is also referred to as grid side for grid-connected inverters. The rated power of single-phase inverters in Germany is 4.6 kW. The PV panels are part of the overall PV system and are typically arranged in strings that provide the power $p_{\mathrm{PV}}(t)$ by the current $i_{\mathrm{PV}}(t)$ and the voltage $u_{\mathrm{PV}}(t)$ that flows into the inverter clamps on the application side. The inverter can also be categorized into hardware components and software components. The vector of the switching signals $\mathbf{sws}(t)$ for the switches at the inverter bridge are the output signal of the software and the input signal of the hardware.

2.1.1 Hardware components

A scheme of the hardware components is depicted in figure 2.2.

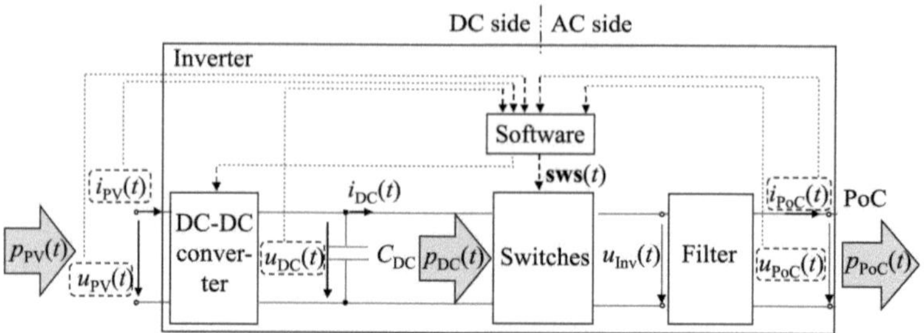

Figure 2.2: General scheme of the hardware components of a photovoltaic inverter.

Switches

In general, insulated gate bipolar transistors (IGBTs) are chosen as switching devices. The IGBTs can be arranged in different topologies, e.g. H5, REFU etc. [36]. Compared to the H4 topology, also known as the full bridge, these other topologies enable to mitigate the currents in zero-voltage states of the switches by making use of additional switches. As a consequence, the power losses can be reduced at the cost of more switches [36]. The topology of the switches is a manufacturer's choice and can differ between the individual designs. However, the voltage spectra at the AC-side terminals remain similar for the known topologies in the frequency range up to 2 kHz as they are all based on the unipolar modulation scheme.

The switching frequency is typically between 5 kHz and 10 kHz which results in an effective switching frequency between 10 kHz and 20 kHz.

DC-link capacitance

The DC-link capacitance C_{DC} consists typically of multiple capacitors that are arranged in parallel and in series. The purpose of the DC-link capacitance is to ensure a DC-link voltage larger than the maximum voltage at the PoC to enable a power flow direction from the DC side to the AC side. Furthermore, the DC-link capacitance needs to buffer the pulsating power of the DC-AC power conversion, i.e. the power flow from the DC side to the AC side is intermittent in single-phase inverters.

In practice, the DC-link capacitance is designed to ensure that the maximum value $\hat{u}_{DC\,rip}$ of the ripple component $u_{DC\,rip}(t)$ which represents the distortion of $u_{DC}(t)$, is below a predefined value, e.g. 5 %, compared to the maximum value of $\hat{u}_{DC}$. In comparison, modern design approaches allow values in the range of 25 %. However, this requires a faster DC-link voltage control and causes other challenges that complicate the controller design. Furthermore, larger ripple voltages on the capacitance cause a higher stress on the DC-link capacitors and reduce their lifetime. On the other hand, the costs for the capacitors increase if a design demands a larger DC-link capacitance.

AC-side filter circuit

The AC-side filter circuit mitigates the distortion of the voltage at the PoC compared to the voltage spectrum at the inverter bridge, i.e. $u_{\mathrm{Inv}}(t)$. Different topologies are applied by individual manufacturers, such as LC-, LCL- or $LCLC$-topologies [36], for reasons of cost-efficiency and weight. The drawback is the increasing number of poles in the transfer functions of the AC-side filter circuit and the overall inverter and thus a possibly large impact on the operation, e.g. a higher risk to cause resonance problems and instabilites.

Further hardware components

While the previously introduced hardware components can be found in all inverters that are known to the author, further components such as a DC-DC converter on the DC side, but also high frequency (HF) transformers are sometimes implemented.
The switching frequency of the DC-DC converter that is above the switching frequency of the IGBTs of the inverter bridge, is consequently chosen far above the bandwidth of the control, so that an interference can be excluded and has not been reported to have an impact on the current distortion at the PoC in the frequency range up to 2 kHz or as an issue for harmonic instabilities.

2.1.2 Software components

The software components include the control, furthermore auxiliary services and protection algorithms. An overview of the control as the main part of the software is depicted in figure 2.3.

Grid synchronization

For first implementations, simple zero-crossing detection has been implemented to synchronize the control reference signals with the power frequency. Nowadays, the grid synchronization of grid-connected low power inverters is based on grid-following algorithms. Devices with a grid-following synchronization algorithm shape the waveform of their current response, i.e. $i_{\mathrm{PoC}}(t)$, based on the voltage characteristics at the PoC, i.e. $u_{\mathrm{PoC}}(t)$, while in practice, also grid-supporting, grid-forming and grid-feeding algorithms can be

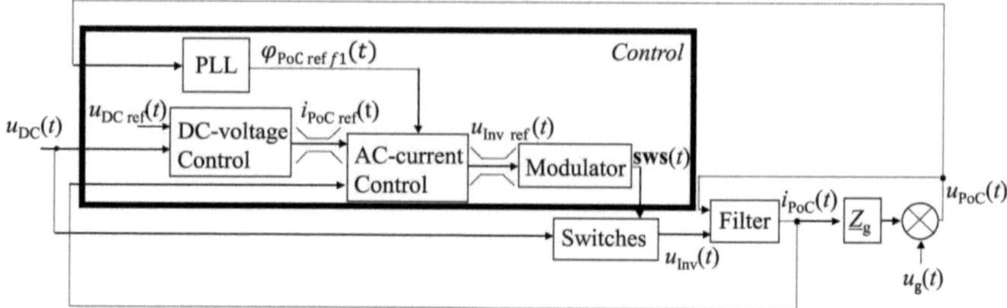

Figure 2.3: Overview of the control.

found in other grid-connected devices [37].

For distorted voltages at the PoC with multiple zero-crossings within one half-period, the zero-crossing detection does not work properly. Therefore, the state of the art grid synchronization is based on PLL algorithms, nowadays. The PLL itself can be assigned to linear control algorithms and adjusts the phase reference $\varphi_{\text{PoC ref}\,U\,f1}$ to the measured input signal, i.e. for inverters the AC-side voltage. However, the use of PLL algorithms in single-phase inverters requires the generation of an artificial orthogonal signal, which is typically created by an orthogonal signal generator (OSG). The OSG can simply delay the input signal by 90° rather called the T4-delay (T4D) with respect to an assumed, fixed power frequency, or can make use of filter techniques, to delay the input signal adaptively. The T4D PLL has to be considered nonlinear if the assumed power frequency and the actual power frequency differ significantly. However, in practice the deviation from 50 Hz is typically below 0.5 Hz, e.g. if the EN 50160 [38] applies, hence the impact of this nonlinearity is negligible compared to the studied harmonic amplitudes resulting from other effects and measurement uncertainties. Furthermore, the T4D is not applied very often. Typically, the OSG is based on filter techniques that can be linearized for harmonic small-signal studies.

DC-link voltage control

The DC-link voltage control is typically implemented with a proportional integral (PI) controller and regulates the DC-link voltage to be above the maximum of the voltage at the PoC $u_{\text{PoC}}(t)$, i.e. the supply voltage. Reference values for the DC-link voltage are between 400 V and 800 V.

AC-current control

The AC-current control is typically implemented with a proportional resonant (PR) controller but can also be implemented with a PI controller. In theory, the phase angle $\varphi_{\mathrm{PoC}\,f1}$ at power frequency between the voltage u_{PoC} and the current i_{PoC} can be set by the AC-current control. However, for low power PV inverters for rooftop applications as studied in the presented work, all performed measurements on commercially available inverters for PV applications with a rated power below 10 kW measured inverters including the test measurements on three-phase inverters have shown a phase angle of 0 ° for $\varphi_{\mathrm{PoC}\,f1}$. Consequently, the analysis of the presented work in the following assumes a phase angle of 0 ° as well. The new code of practice in Germany, i.e. the VDE-AR-N 4105 [39], demands a grid-supply by reactive power if the root mean square (RMS) value of the voltage at the PoC drops too low. For this work, the fundamental frequency of the background voltage, i.e. the power frequency, is considered with 50 Hz and an RMS value of 230 V respectively as the topic of power balance between active and reactive power is not part of the harmonic study objective.

Modulator

For single-phase inverters and converters using pulse width modulation (PWM), the resulting spectra are explained in [40] based on the double Fourier integral analysis. For real devices, the unipolar asymmetric regular sampled PWM represents the state of the art compared to simplified representations, e.g. naturally sampled modulation algorithms. The time-domain solution of the line-neutral voltage at the inverter bridge $u_{\mathrm{Inv}}(t)$ is formulated with

$$
\begin{aligned}
u_{\mathrm{Inv}}(t) = & \frac{4u_{\mathrm{DC\,av}}}{\pi}\sum_{n=1}^{\infty}\frac{J_n\left(n\frac{f_1}{f_{\mathrm{car}}}\frac{\pi}{2}m_{\mathrm{a}}\right)}{n\frac{f_1}{f_{\mathrm{car}}}}\sin\left(n\frac{\pi}{2}\right)\cos(2n\pi f_1 t)+ \\
& \frac{4u_{\mathrm{DC\,av}}}{\pi}\sum_{m=1}^{\infty}\sum_{k=-\infty}^{\infty}\frac{1}{q}J_{2k-1}\left(q\frac{\pi}{2}m_{\mathrm{a}}\right)\cos([m+1-k]\pi)\cdot \\
& \cos(4m\pi f_{\mathrm{car}}t+[2k-1]2\pi f_1 t)
\end{aligned}
\tag{2.1}
$$

for the DC-link voltage $u_{\mathrm{DC\,av}}$ that is averaged over a period of the power frequency f_1, i.e. the fundamental frequency of the AC-side voltage at the PoC $u_{\mathrm{PoC}}(t)$, the carrier frequency of the modulation f_{car}, the incremental parameters n and m, the modulation index m_{a} with

$$
m_{\mathrm{a}} = \frac{\hat{u}_{\mathrm{PoC}}(f_1)}{u_{\mathrm{DC\,av}}},
\tag{2.2}
$$

the modulation parameter q with

$$q = 2m + [2n - 1]\frac{f_1}{f_{\text{car}}} \tag{2.3}$$

and the Bessel functions J of first kind at order n with

$$J_n(x) = \sum_{k=0}^{\infty} \frac{(n-1)^k (\frac{x}{2})^{n+2k}}{k!\Gamma(n+k+1)} \tag{2.4}$$

and the Gamma function Γ with

$$\Gamma(x) = \int_0^{\infty} e^{-t} t^{x-1} \mathrm{d}t. \tag{2.5}$$

The simplification of the DC-link voltage being averaged, i.e. $u_{\mathrm{DC\,av}}$, and the AC-side voltage containing only f_1 represents still the state of the art. For device compliance tests of harmonic emission limits, only sinusoidal (non-distorted) wave forms of $u_{\mathrm{PoC}}(t)$ are used so that for standard PE design studies only the sinusoidal supply voltage is of interest. In real systems, a distorted signal that includes voltage frequency components at frequencies other than f_1 is typical. Detailed white-box models can inherently represent the current response of the inverter in terms of a distorted current by also considering a distortion in the voltage. However, the analytic solution has not been studied for distorted voltages, since for manufacturers, it is sufficient to comply with the regulations and standards whereas white-box models are typically not available for network operators.

Protection and auxiliary services

Required and optional additional algorithms are implemented, e.g. to protect the inverter and to handle abnormal operating conditions. In case of overcurrents and overvoltages, direct shutdowns of the inverter are triggered instantaneously. For the optimal power flow, the maximum power point (MPP) enables an optimal operation of the PV panels by slightly varying the PV string voltage attached to the DC-side clamps of the inverter. Tracking the power based on the current flowing into the clamps on the DC side of the inverter, the maximum operating power can be achieved. Furthermore, anti-islanding (AI) detection is found as well as algorithms that detect a minimum power flow on the DC side to start the inverter operation and enable the power injection into the LV network. These algorithms take time to evaluate the signals and react with a delay.
The maximum power point tracker (MPPT) affects the current spectrum but not in the harmonic range, i.e. the frequency range on which the following work focuses on. The time constants of the MPPT are in the range of some hundred milliseconds and thus only affect frequencies below 50 Hz.

All other services affect the status of the operation, i.e. on or off, but do not affect the manner of the operation, e.g. the distortion of voltages and currents, since they are typically running in the background of the software signal processing of the inverter. Consequently, none of the protection and auxiliary services affect the current distortion in the harmonic range.

2.2 Modeling approaches

Modeling approaches can be sorted according to different criteria. Firstly, the models can be categorized according to the model domain, e.g. frequency domain, time domain or wavelet (frequency-time) domain. Secondly, the implementation type can account as another criterion, e.g. by analytic descriptions, signal flow diagrams and function block diagrams as well as electric circuit simulations.
However, when relating to commercially available devices and simulations, a relevant aspect is the available knowledge about a specific inverter that is either known, generically implied or partially available for some inverter components, or unknown and has to be measured. With regard to a knowledge-based classification, three modeling approaches can be defined. These three approaches are called the white-box, the grey-box and the black-box approach.

2.2.1 White-box

The development of white-box models requires the highest degree of knowledge. The topology of all components, i.e. software and hardware, and the respective parameters have to be known.

Switched model

The most elaborate white-box model is the switched model that represents a device entirely including the nonlinear characteristics of the control and of the semiconductor switches. In practice, a simplification of the switched model can be suitable in terms of a linearization or the neglection of component characteristics that are not relevant with regard to the objective of an individual study.

Linearization

Since linearity is a system characteristic, the relation between an input signal and an output signal can be linear, but not the signal itself. By definition, linearity entails the applicability of the principle of superposition. With regard to the input and the output signal of the system, this can imply a direct proportionality but also an indirect proportionality.

Operating point

An operating point is called the set of impact factors that define the operation of the inverter. This includes the internal impact factors as well as external impact factors, e.g. the input signals. Inverters are not linear in general but when studying their small-signal characteristics, they can be linearized around a specific operating point. More details will be given in the following of this chapter as well as in chapter 3.

Reference point

The operating point with a specific set of input signals that is chosen for the model development is called the reference point.

Small-signal model

The small-signal model represents the linearized inverter representation for one operating point. The hardware components are considered linear in the frequency range up to 2 kHz while the control is typically linearized, e.g. the PLL, the DC-link control and the AC-side current control.

Linear time periodic model

The small-signal model of the inverter can be represented by a linear time-periodic (LTP) model with the general state space description in time domain in terms of

$$\dot{\boldsymbol{x}}(t) = \boldsymbol{A}(t)\boldsymbol{x}(t) + \boldsymbol{B}(t)\boldsymbol{u}(t) \tag{2.6a}$$

$$\boldsymbol{y}(t) = \boldsymbol{C}(t)\boldsymbol{x}(t) + \boldsymbol{D}(t)\boldsymbol{u}(t) \tag{2.6b}$$

with the input vector $\boldsymbol{u}$, the output vector $\boldsymbol{y}$, the state vector $\boldsymbol{x}(t)$ and

$$\dot{\boldsymbol{x}}(t) = \frac{\mathrm{d}}{\mathrm{d}t}\boldsymbol{x}(t) \tag{2.7}$$

for the real-value system parameters $\boldsymbol{A}(t)$, $\boldsymbol{B}(t)$, $\boldsymbol{C}(t)$ and $\boldsymbol{D}(t)$ and the periodicity T by means of

$$\boldsymbol{A}(t) = \boldsymbol{A}(t+T). \tag{2.8}$$

This small-signal model can be separated into a constant part, i.e. the linear time-invariant (LTI) characteristics, and a periodically changing part, i.e. the LTP characteristics [41].

Periodicity

The periodic changes of the system itself lead to new frequency components by convolution in frequency domain with the input signal. The resulting frequency components that differ from the excited frequency are called frequency coupling components. Frequency coupling components are generated inside the inverter when considering a two-pole representation, or rather a single port that is connected to an electric network at the AC side.

Harmonic state space model

Following the Floquet theorem, the time-domain representation of the LTP state space model can be transformed into frequency domain by using the monodromy matrix [41] and is called the harmonic state space (HSS) model. The expansion of (2.6) by means of the complex Fourier series with

$$\boldsymbol{A}(t) = \sum_{n=0}^{\infty} \boldsymbol{A}_n \, \mathrm{e}^{jn2\pi f_1 t}, n \in \mathbb{N} \tag{2.9}$$

and following for $\boldsymbol{B}(t)$, $\boldsymbol{C}(t)$ and $\boldsymbol{D}(t)$ is also applied to the input vector in terms of

$$\boldsymbol{u}(t) = \sum_{n=0}^{\infty} \boldsymbol{u}_n \, \mathrm{e}^{jn2\pi f_1 t} \tag{2.10}$$

and respectively for $\boldsymbol{x}(t)$ and $\boldsymbol{y}(t)$. For the white-box analysis, the appropriate matrices can be calculated based on the full model description represented by the individual component topologies and the respective parameters. Since the resulting Toeplitz matrix contains an infinite number of elements, it is appropriate to reduce the matrix by truncation considering the frequency range of interest with K being the order of truncation. With regard to the input and output, it follows

$$\boldsymbol{u} = [u_0(0f_1), u_1(1f_1), u_2(2f_1)..., u_K(Kf_1)]^{\mathrm{T}} \tag{2.11}$$

and for $\boldsymbol{x}$ and $\boldsymbol{y}$ respectively. It is possible to derive the HSS model by means of the harmonic transfer functions (HTF) H_k that are expressed in the matrix form

$$\boldsymbol{H} = \begin{bmatrix} \underline{H}_0(0f_1) & \underline{H}_0(1f_1) & \underline{H}_0(2f_1) & \dots & \underline{H}_0(Kf_1) \\ \underline{H}_1(0f_1) & \underline{H}_1(1f_1) & \underline{H}_1(2f_1) & \dots & \underline{H}_1(Kf_1) \\ \underline{H}_2(0f_1) & \underline{H}_2(1f_1) & \underline{H}_2(2f_1) & \dots & \underline{H}_2(Kf_1) \\ \vdots & & \vdots & \vdots & \ddots \\ \underline{H}_K(0f_1) & \underline{H}_K(1f_1) & \underline{H}_K(2f_1) & \dots & \underline{H}_K(Kf_1) \end{bmatrix}. \tag{2.12}$$

For the analytic development of $\boldsymbol{H}$, the individual inverter components are related to the specific matrix elements of relevance so that the elements of $\boldsymbol{H}$ can be represented as terms in a first step but also be aggregated and abstracted to complex values.
The main diagonal elements of $\boldsymbol{H}$ represent the LTI characteristics and determine the system output signal, e.g. the inverter current, at the same frequencies as the input signals for

$$\boldsymbol{y} = \boldsymbol{H}\boldsymbol{u}. \tag{2.13}$$

The off-diagonal elements introduce frequency responses that differ from the related frequency of the input signal, i.e. the frequency coupling components. If the proportionality between an input signal and an output signal remains the same the system is linear, also in presence of frequency coupling components, and can be represented as in (2.12).
To analyze, if the system responds linearly with regard to the input and output signals, the origin of frequency coupling components is consequently also of importance. In [42], different effects of coupled and decoupled models in PE systems are described. The implemented model components are categorized into mirror frequency decoupled (MFD) components, e.g. linear passive R-, L-, C-components and control algorithms in $\alpha\beta$-frame, and mirror frequency coupled (MFC) components, e.g. nonlinear control algorithms. Though this analysis is performed in dq-frame, it demonstrates the individual component impacts on the frequency couplings and the frequency spectrum of the current. However, it is important to furthermore distinguish between single-phase and three-phase topologies, specifically when analyzing the DC-link voltage or the PLL. In [43], the spectrum of the DC-link voltage of single-phase inverters is qualitatively described in terms of frequency components at $f_1 \pm f_1$, $2f_1 \pm f_1$ and $4f_1 \pm f_1$. A quantitative description has been introduced in [40] as annotated in the previous section 2.1.2.

Averaging

If an inverter model is averaged over the power frequency f_1, the model is reduced to its LTI characteristics.

Neglection

Next to averaging, different effects are neglected in various studies based on the study objective and the studied devices to simplify the study analysis.

1. Frequency coupling components in the voltages and currents are sometimes neglected, e.g. due to missing information or the lack of knowledge.

2. The frequency range is reduced to a specific range, e.g. often studies are performed in the subharmonic, harmonic or supraharmonic frequency range.

3. Dynamics and individual device component characteristics are also often neglected. Typically, the impact of the switching of the IGBTs is neglected for harmonic studies.

4. The impact of the DC-link capacitance on the voltage is often neglected. For harmonic stability studies, the DC-link voltage cannot be represented as a constant voltage source if DC-link voltage dynamics have to be included in the analysis. Compared to a capacitance, an uncontrolled voltage source is always stable and can provide infinite power to the system. Not only the DC-link voltage itself is inherently stable in those simulations but also the impact of the DC-voltage control is neglected since the voltage source can always provide sufficient power while in practice, the capacitor size is a crucial parameter in interaction with the DC-link voltage controller for the harmonic stability of an inverter.

2.2.2 Grey-box

Compared to the development of white-box models, grey-box models include as much information as possible though not all details are available. Regarding the unavailable knowledge about components and component equivalent models, the grey-box modeling approaches can be further distinguished into methods to identify specific component topologies and methods to identify specific parameters of a known component topology.

Component identification

Often, the general topology of a device is known due to its operating principle. An approach that has not been found in literature but has been looked into for the development of suitable inverter models in the harmonic frequency range is the component identification. If equivalent models for required inverter components can be found and identified, there is a possibility to develop an overall equivalent model that can include as much information about the inverter as possible to be as detailed as possible.
Beginning with the hardware, the AC-side filter circuit as one of the main inverter components with regard to the interaction of harmonic currents at the PoC has been studied and a method for its frequency domain equivalent model has been proposed in [8] by the author and is disclosed in appendix D.1.
As current state of the art, it is not possible to identify the topology of the control, e.g. cascading of multiple controlers, due to the large number of possible implementations that show a similar behavior for different operating points. One of the main challenges is the cascaded structure of the control implementation that makes a large range of implementations possible. Respective measurement-based methods to separate the different controllers or identify an equivalent model have not been developed. First ideas on the

identification of the control are annotated in appendix D.3 but require a more detailed study.

Parameter identification

To identify the parameters of individual device components, different methods have been proposed while general approaches are well established since a long time, e.g. [44]. More recently, [45] has applied a non-invasive method to identify the circuit and control parameters of a voltage source converter (VSC) though this requires detailed knowledge about the structure of the control and also of the hardware configuration that is typically not available for commercially available inverters.
In the later of this work, the DC-link capacitance is identified as the main origin of LTP frequency coupling components in the current at the PoC. An identification of the DC-link capacitance has been disclosed in [9] by the author and is presented in appendix D.2 that enables to consider the aggregated capacitor characteristics of the DC-link.
Together with the identified AC-side filter circuit and an aggregated representation of the control, an overall grey-box model can be developed in the future to enable dynamic time-domain studies.

2.2.3 Black-box

Time-domain models ([46, 47]) and frequency-domain models have been developed for black-box analyses and simulations. A very simple black-box model is an uncontrolled current source that represents a current spectrum measured at one operating point. If this black-box model is used for simulations, deviations from the reference point chosen for the current source parametrization are neglected. To include dependencies on the voltage at the PoC, two further models have been developed in frequency domain, namely the decoupled Norton model and the more advanced coupled Norton model. Both models are small-signal models that apply linearization around the reference point and are developed for harmonic studies.

Decoupled Norton model

The decoupled Norton model represents the inverter by its LTI characteristics and is depicted in figure 2.4. The uncontrolled current source simulates the current that has been measured at the reference point. The admittance $\underline{Y}_{\mathrm{Inv}}$ takes the deviations from the applied reference voltage at the PoC into account. If the inverter is identified at a sinusoidal reference voltage at the PoC, any frequency component at any operating point besides the power frequency will represent a deviation from the reference voltage and is

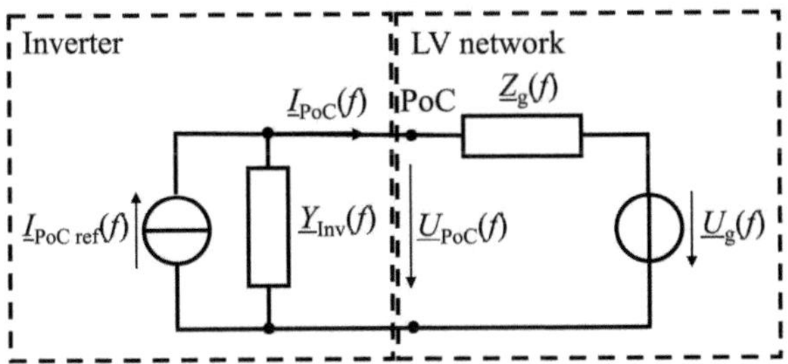

Figure 2.4: Decoupled Norton model of an inverter and decoupled Thevenin model of the LV network based on [11].

applied to $\underline{Y}_{\mathrm{Inv}}$ as shown in figure 2.4. The decoupled Norton model as shown in figure 2.4 represents a simplification of the small-signal characteristics of the inverter.

Coupled Norton Model

A more advanced model than the decoupled Norton model exists and is called the coupled Norton model. The coupled Norton model considers also the multiplications of signals in time domain, e.g. the switching of the IGBTs and the voltage dynamics of the DC-link. This multiplication of signals in time domain or rather a convolution in frequency domain generates new frequency components in the voltage spectrum, e.g. $\underline{U}_{\mathrm{Inv}}$ according to (2.1). While time-periodic system theory has already been introduced in [41], the resulting models of nonlinear electric loads have been called admittance frequency coupling matrices [48] and been alternatively renamed to harmonically coupled admittance matrices with regard to AC-DC converters [49]. As the classical admittance according to fundamentals of electrical engineering is constant and not periodic, the author calls the admittance frequency coupling matrix or rather harmonically coupled admittance matrix (HAM) simply a frequency coupling matrix (FCM) $\boldsymbol{Y}_{\mathbf{Inv}}$.

The admittance $\underline{Y}_{\mathrm{Inv}}$ reflects the current characteristics for voltage deviations $\Delta\underline{U}_{\mathrm{PoC}}$ from the reference point, i.e. in figure 2.5 for a sinusoidal reference voltage thus $\Delta\underline{U}_{\mathrm{PoC}}$ is equal to $\underline{U}_{\mathrm{PoC}}$ for voltages above power frequency, if the voltage frequency f_U and the current frequency f_I are the same. The equivalent circuit model is represented at the frequencies f_I. As mentioned previously, voltage frequency components can affect currents at other frequencies due to a convolution in frequency domain. These so-called frequency coupling components are represented by the voltage-controlled current source for voltage frequencies that differ from the current frequencies. The current at the PoC

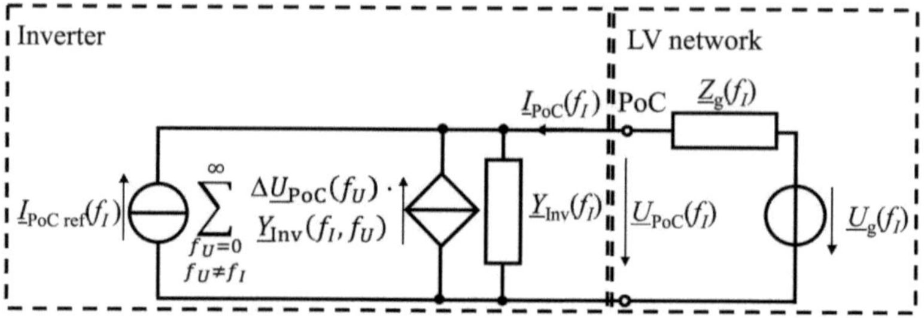

Figure 2.5: Coupled Norton model of an inverter and decoupled Thevenin model of the LV network based on [16].

can be described based on Kirchhoff's first law with

$$\underline{I}_{\mathrm{PoC}}(f_I) = \underline{I}_{\mathrm{PoC\,ref}}(f_I) + \underline{Y}_{\mathrm{Inv}}(f_I)\underline{U}_{\mathrm{PoC}}(f_I) + \sum_{\substack{f_U=0\\ f_U\neq f_I}}^{\infty} \Delta\underline{U}_{\mathrm{PoC}}(f_U)\underline{Y}_{\mathrm{Inv}}(f_I, f_U). \tag{2.14}$$

Measurement-based parametrization

To identify an LTP model, e.g. a coupled Norton model, [41] has proposed to perform a frequency sweep (FS) over a single-frequent input signal. However, the first application was not on PE devices but on helicopter rotors [50] and systems with parametric excitation. Consequently, the method has been adapted and called the fingerprint for the application on PE devices [51]. In addition to the single-frequent sweep of the input signal, which is for the fingerprint the voltage at the PoC, an additional voltage component at power frequency is applied during the fingerprint to enable an operation of the device under test, e.g. an inverter.

When appliying the fingerprint, firstly, a reference voltage $\underline{U}_{\mathrm{PoC\,ref}}$ is applied at the PoC and the respective multi-frequent current response $\underline{I}_{\mathrm{PoC\,ref}}$ is measured. Afterwards, an additional sweep component in the frequency spectrum of the voltage is applied and the current responses are measured for all measurement points. The elements of the FCM can be calculated by

$$\underline{Y}_{\mathrm{Inv}}(f_I, f_U) = \frac{\underline{I}_{\mathrm{PoC}\,i}(f_I) - \underline{I}_{\mathrm{PoC\,ref}}(f_I)}{\underline{U}_{\mathrm{PoC}\,i}(f_U) - \underline{U}_{\mathrm{PoC\,ref}}(f_U)}. \tag{2.15}$$

It is possible to vary the fingerprint by also testing different amplitudes and phases to ensure linearity of the device under test (DUT). In practice, testing a set of phase angles for all respective amplitudes and frequencies factorizes the measurement-based parametrization and leads to a long overall measurement duration. The later of this work studies the linearity of the small-signal characteristics of inverters. In literature, an adaption of the FS has been proposed in [52] where the phase angle is not changed stepwise to avoid waiting until a steady state is reached for the new measurement point. Instead, it proposes to change the phase angle continuously to reduce the measurement duration. However, this is only an advantage for devices that are dependent on the phase angle of the voltage harmonics at the PoC and can still be represented by FCMs.

It is possible to develop an averaged model by averaging over the phase angle and the amplitude but the advantages with regard to practical reasons are not very apparent.

First studies of the impact of the operating power on the small-signal characteristics have been presented in [53] with respect to subgroup total harmonic distortion (THDS). Individual frequency components at even order harmonics two four and six and odd order harmonics up to order 19 are depicted in [54] but the assessment focuses on the total harmonic distortion (THD) thus they are not suitable for a frequency-dependent analysis that is required for the stability analysis.

Linearity assessment

To assess the linearity of PE devices, indices have been introduced, namely the Fourier descriptors [55]. Fourier descriptors are used with regard to the previously introduced coupled Norton model. The representation of the voltage-current relations and the resulting impedance or rather admittance characteristics resemble circles and ellipsoids when plotted on polar plots. The similarities of the figure on the polar plot in comparison to ideal circles can be expressed be the Fourier descriptors. However, the Fourier descriptors assess the linearity of the voltage-current relations when comparing to individually selected, single frequencies but do not make an assessment of superimposed voltage and current frequencies possible, i.e. an overall model linearity.

Nonlinear models

To overcome the constraints of frequency domain models, time-domain models with a nonlinear structure have been studied in the past. The Hammerstein-Wiener (HW) model has been studied firstly for inverters in [56] and later also by the author in [24, 25]. In addition, the feasibility of artificial neural networks (ANNs) has been studied by the author in [7]. It can be concluded that the Coupled Norton model does not perform worse than the studied nonlinear modeling structures but is easier to parametrize with a simpler model structure. Furthermore, these newly studied models require more elaborate studies of the general feasibility for simulations of inverter instabilities and the parametrization of these models.

2.3 Stability assessment

2.3.1 Power system stability

From a general perspective, [57] describes the power system stability as the property of the system to operate at an equilibrium point or to restore the operation at an equilibrium point after a subjection to a disturbance. The power system stability has been classified into rotor angle stability and voltage stability. In addition, small-signal stability, transient stability, mid- and long-term stability as well as the classification into small and large disturbance signals have been introduced by [57]:
The rotor angle stability is described in relation to the sychronism of synchronous machines in power systems. If an interconnected synchronous machine runs temporarily faster than another interconnected synchronous machine, the load of the slower synchronous machine will be transferred partly to the faster synchronous machine thus the angular seperation between these two synchronous machines is reduced. If the angular separation of one synchronous machine is too large in relation to the rest of the power system, this synchronous machine loses its synchronism and thus the ability to restore its equilibrium point and becomes instable.
The voltage stability relates to the buses in a power system and requires an acceptable voltage at all buses. The voltage stability is consequently refering to a demand in reactive power.
The small-signal stability describes the ability of a system to maintain the equilibrium point or to restore it when being subjected to small disturbances. On the other hand, the transient stability describes the ability of a system with regard to large disturbance. After the occurence of such large disturbances, a stable system often restores the operation at a new equilibrium point.
The CIGRE Study Committee (SC) C4 and the IEEE Power System Dynamic Performance Committee have recently introduced an adapted definition by classifying into frequency stability, voltage stability, rotor angle stability, converter driven stability and resonance stability [58]. The following of this work is related to the converter driven stability.
The author of this work proposes rather to refrain especially from the resonance stability as it uses the term stability. The so-called resonance stability is solely a resonance phenomenon including the physical characterstics of resonances, e.g. passivity. It appears to be misleading to speak of a stability as the so-called resonance stability is not related to classical stability definitions, e.g. an amplification of the input signal that can result from a positve feedback.
Nevertheless, the resonance phenomenon and immunity issues present a challenge in practice that cause devices in energy networks to shut down. However, this requires a contin-

uous excitation as the damping is present but neglibly low and the network impedance is passive. A more general view on the stable operation and the device immunity are adressed in section 5.6 and section 5.7.

2.3.2 Inverter stability

The stability of the inverter can be categorized in relation to the signal characteristics of the voltage at the PoC, e.g. into the static stability by means of the small-signal stability and the transient stability as large-signal stability. Large-signal analyses refer to changes that include changes at power frequency and consequently allow a global stability assessment.

2.3.3 Harmonic stability

The harmonic stability is categorized as a small-signal stability and describes the interaction of the inverter control, the AC-side filter circuit and the network impedance [30]. If the inverter operates at an equilibrium point due to this interaction or can restore the equilbirum point after small-signal disturbances, the inverter operates stable. The frequency range of the analysis is above 50 Hz up to 2 kHz or rather the 40th order of the power frequency. The 40th order of the power frequency is typical for harmonic assessments according to IEC 61000 standards, e.g. the IEC 61000-3-2 [59], but also with regard to the LV network, e.g. in Europe the EN 50160. For this work, it is important that the bandwidth of the control, which is smaller than 1 kHz, is covered by the harmonic frequency range [10].

Eigenvalue analysis

A common approach to analyze the harmonic stability of inverters is the Eigenvalue analysis. The development of the respective small-signal representations, e.g. achieved by linearization, requires knowledge about the internal design of the inverter, which is typically only available to manufacturers and research facilities with access to the software and hardware parameters. Though the final small-signal model can be abstracted to a black-box representation, a reverse development, e.g. of a measurement-based identification of the black-box model and a recalculation of the Eigenvalues based on fitting and optimization algorithms is not known as of yet.
Related to the Eigenvalues, a respective pole-zero mapping can be performed to evaluate changes of the poles, e.g. of the detailed and parameter-based HSS, in the complex s-plane related to changes of the inverter parameters, e.g. [43]. In general when applying pole-zero mapping, a system is identified as instable if the poles of the HTFs from (2.12)

are in the right half plane (RHP) by means of

$$H_k(s + j2n\pi f) = \frac{N_n(s + j2n\pi f)}{D_n(s + j2n\pi f)} \tag{2.16}$$

with the general representation of the nominator by N and the denominator D, which has been demonstrated in [60].

Nyquist stability criterion

In practice and for the later of this work, certain restrictions narrow down the applicable stability approaches based on the available data. For commercial inverters, the design parameters are unknown. The data that is used for the assessment can only be generated based on measurements. The achieved model, e.g. the FCM, is typically discrete and abstract, i.e. represented by complex values instead of terms based on the inverter parameters.

The Nyquist stability criterion enables a stability analysis based on the Nyquist plot. The Nyquist plot can be derived by plotting the open loop transfer function on the complex plane, i.e. Cartesian coordinates. The stability can be assessed for linear and linearizable systems by studying the encirclement of point (-1, j0). For simple system representations, such as the public LV network and its interaction with the inverter in terms of the respective impedance representations, the impedance-based stability can be assessed. Simplifying the general Nyquist criterion, it can be stated that if the Nyquist plot of the open loop transfer function does not encircle the point (-1, j0) at all, the system is stable. The formal analysis by the Nyquist criterion can be described by introducing two criteria: the amplitude criterion and the phase margin criterion. Considering the point (-1, j0), the magnitude of the open loop system transfer function is one. With regard to the open loop transfer function of the entire system, i.e. the LV network and the inverter, the relevant terms are the network impedance and the inverter impedance. The magnitude of the open loop transfer function relates to the ratio of the LV network impedance and the inverter impedance [61], which is shown in more detail in the later chapter 5.

If the ratio is one, i.e. the magnitudes of the inverter and the network impedance intersect when being plotted over the frequency (amplitude criterion), the phase margin φ_{PM} must fulfill the criterion for the phase angle of the network impedance $\varphi_{\mathrm{g}\,Z}$ and the phase angle of the inverter impedance $\varphi_{\mathrm{Inv}\,Z}$ with

$$\varphi_{PM} = 180° - \varphi_{\mathrm{g}\,Z}(f) + \varphi_{\mathrm{Inv}\,Z}(f) > 0°. \tag{2.17}$$

As an explanation, both criteria in combination analyze if the trajectory of the open loop transfer functions fulfill the Nyquist criterion. The inverter impedance can be calculated analytically based on knowledge about the parameters or by performing a measurement-based identification by following the principle of the FS as described in section 2.2.3.

2.4 Research gaps and research potential

Open research gaps in relation to the state of the art have been identified at the beginning of this work. Many of these open research gaps are addressed and solved in the later of this work while others are still left to future work but will partly be mentioned in the later chapters.

2.4.1 Design

The inverter designs in industry are typically developed based on internal company design rules. These rules and internal parameter tables enable the manufacturers to design a certain number of devices with a rather similar design. The objective of the design development is the functionality in the field and not necessarily the most detailed understanding of all signal interactions. To chose the most suitable component design and to improve the design rules, a holistic understanding of possible and present implementations of the devices is required. To compare available devices, a modular modeling approach can be developed that compares individual components and at the same time enables to generate a large database of device models to study the individual characteristics of different makes of devices in terms of their individual component designs. This modular approach has been developed and implemented for the following work and is disclosed in [2] and described in A.1.

2.4.2 Linearity analysis

In practice, the black-box representation of an inverter is often introduced as a small-signal model. However, the limitations of the small-signal model, its dependencies and the operating point for which the individual small-signal model is suitable are typically not disclosed.

A linearity framework is required that describes the linearities of a device with regard to the external signals, i.e. the input-output signal characteristics and the voltage-current relation of the inverter at the PoC and consequently the limitations of an individual small-signal model.

Detailed white-box studies for input-output signal characteristics of selected components, e.g. the PLL, are available. However, for some components, it is challenging to analytically describe their characteristics and requires sophisticated knowledge. Consequently, not all analytic component descriptions for all components and individual designs are available.

In practice, simplification measures, e.g. linearization, are applied but the accuracy of the models and the limits of the linearity are not disclosed.
To understand the voltage-current relation, not only of one specific device design but a representative number of different device designs, general background knowledge is required in terms of the linearities of the implemented components. Knowledge of the linearities of a device is of importance to identify the sweep parameters of the frequency sweep and is presented in chapter 3.

2.4.3 Measurement-based identification

The knowledge about the linearity of the small-signal models with regard to the input signals is required to develop a suitable measurement-based identification method. In detail, this implies the identification of relevant impact factors.
As the following work presents inherently the importance of the frequency dependency, [53] and [54] are interesting to start off with but are not sufficient to perform stability analyses. In detail, the studies are too generic and do not disclose the relevance of possible input factors.
More general, various studies have presented case studies for a set of measurements but no holistic framework has been developed. Nevertheless, such a structured measurement framework is required for a reliable identification of inverter dependencies and a definition of the dependencies on the stability. The reduction of measurement points by neglecting irrelevant changes of the input signals is required. A parameter sweep for all possible parameters is not applicable to a large number of different devices of one device class. To optimize the test signal by applying the principle of superpostion, a first proposal has been disclosed by the author in [11] and is presented in the later chapter 4 in section 4.3. In general, knowing the device characteristics, e.g. the independent, linear and nonlinear dependencies on the input signals, the measurement-based identification methods can be designed appropriately. Consequently, the relevant impact factors have to be identified to reduce the measurement points of a generally applicable identification method.
A future, alternative modeling approach can possibly identify individual components and the device behavior based on measureable signals to apply grey-box modeling techniques. With regard to the black-box modeling approach, the existing FS is too general. Black-box identification methods are available, but it is typically not disclosed how the individual model has been identified. The impact of individual voltage frequency components related to the entire current spectrum has to be studied for each device class, e.g. inverters. Not only the individual frequencies but also the superposition and hence the specific set of frequencies in terms of individual combinations can affect the entire spectrum as known from rectifiers which is why chapter 3 studies the linearity of inverters.
It takes up to multiple days to measure all measurement points and creates hundreds

of gigabytes (GBs) of data. The generated data has to be evaluated which requires a large effort, especially time, but also a large storage capacity. In general, reference point conditions and, even further, the details of the FS regarding the applied voltage spectra (tested frequencies, phase angles and amplitudes) and the applicable measurement uncertainty is not disclosed or verified in many studies.
Therefore, the test signals for the identification, e.g. based on the FS, have to be defined to develop suitable and time-efficient identification methods.

2.4.4 Modeling

Simulation environment

The results and the inverter characteristics in case of simulative studies have to be assignable to the individual inverter design and must not originate from unsuitable solver parameters (time-steps, relative accuracy etc.). To analyze the impact of solvers and further to identify suitable solver settings, a study on different solvers has been performed by the author [5] while the relevant details are disclosed in appendix A.3. For most simulations, it is not suitable to set the maximum solver accuracy because of a too high computational effort. Consequently, the type of the solver and the solver settings have to be chosen based on a defined, feasible accuracy and to avoid convergence issues.

White-Box

White-box models require detailed knowledge about all device parameters that are typically not disclosed by the manufacturers. In addition, manufacturing deviations cause large uncertainties of the individual inverter component parameter values compared to the intended component parameter values, especially for hardware components.
It is necessary to reduce the implementation effort, especially for multi-design studies, i.e. studies that are more general and consider different manufacturer designs. For these studies, a modular modeling approach is pursued by the author and disclosed in appendix A.1.
Since analytic solutions have only been studied for selected components with specific topologies and parameters, the aggregated behavior of a device in terms of the frequency coupling components is still not achieved and studied in chapter 3.
Furthermore, simplifications of white-box models are often performed, e.g. by an uncontrolled voltage source to represent the DC-link capacitance and the remaining PV system components that are connected to the DC-side inverter clamps, e.g. the PV modules etc. These simplifications can lead to neglecting relevant device characteristics that are of importance for the harmonic stability thus knowledge of the physical background of harmonic instabilities is required to apply only proper simplifications.

Grey-Box

The current challenge of grey-box modeling is the missing methods to identify all relevant inverter components. Identification methods and the structure of a new grey-box model have been studied by the author and the results are presented in appendix D, i.e. an identification method of the AC-side filter circuit and the DC-link capacitance based on [8,9]. Still, the identification and a model representation of the control and thus also an overall grey-box model structure of an inverter has not been developed as of yet.

Black-box

Black-box identification methods are available, but it is typically not disclosed how the individual model has been identified. The impact of individual voltage frequency components related to the entire current spectrum have to be studied for each device class, e.g. inverters. Not only the individual frequencies but also the superimposed frequencies and hence the specific set of frequencies in terms of individual combinations can affect the entire spectrum as known from rectifiers which is why chapter 3 studies the linearity of inverters.

State of the art time-domain black-box models are not suitable for stability studies. The requirement for model-based stability studies is the model ability to include dynamic inverter interactions with the LV network at the PoC. Currently developed models are typically transformed from frequency domain into time domain but are static. Furthermore, they converge only for steady system states. Applied solver algorithms have convergence issues when solving harmonic power flow equations for instable operating conditions. A clear formulation of initial conditions and boundary conditions or rather abortion criteria are currently not defined and have to be identified and defined. In addition, the computational effort of dynamic time-domain models is very large. Consequently, there is a need to simplify the degree of detail to reduce the computational effort without losing the important model details that represent the origins of harmonic instabilities. Therefore, future models have to be validated also for system dynamics and not only for steady state.

With regard to future black-box models that are able to handle system dynamics, the model structure of ANNs has been tested by the author to open up the research field of black-box modeling of PE devices in time domain [7]. In detail, the feasibility of ANNs has been studied for simulating dynamic transitions between different steady states. The task to detect dependencies between currents and voltages at the PoC under consideration of different operating points and changes of the operating point is given to the ANN. The ANN can be trained by sets of input, output and validation data to adapt its parameters. However, even for steady state operating points, knowledge of the expected inverter op-

eration and linearities are required for the representation of possible operating states by an ANN. This still requires detailed understanding of the physical operation of inverters. If such detailed knowledge is available, it is likely to also find suitable analytic descriptions of the physical dependencies thus the use of the ANN is not required and the ANN can be replaced by the implementation of the exact physical formulas. Furthermore, the ANN model has not been studied with respect to its feasibility of simulating harmonic instabilities, which still represents an open research field for future studies.
In literature, hybrid modeling approaches are often proposed. The claim is, to develop future frequency-time domain dynamic models that transfer within the simulation between frequency and time domain. Such a hybrid model has not been successfully validated for dynamic black-box modeling of PE devices in general. One of the main challenges is the Heisenberg-Gabor uncertainty principle [62] that has not been overcome. The Heisenberg-Gabor uncertainty principle states that the product of time resolution σ_t and frequency resolution σ_f is constant which can be formulated by

$$\sigma_t \sigma_f \geq \frac{1}{4\pi}. \tag{2.18}$$

This limits hybrid modeling approaches since during simulations, a frequency-time domain dynamic model cannot exist that accurately represents the dynamic behavior and accounts especially for physically instable systems that do not converge. A frequency-time transformation, e.g. a Fourier transform, based on frequency domain models, i.e. a hybrid modelling approach, cannot solve this issue. Any function $\hat{f}$ defined with

$$f \rightarrow \hat{f} := (\langle f | e^{2\pi ikt} \rangle), k \in \mathbb{Z} \tag{2.19}$$

that relates the Fourier coefficients to $f \in L^2([0,1])$ is called a Hilbert space isomorphism as explained by the author in [13], which implies that a single frequency component of a multi-frequent response on an input signal, i.e. the voltage at the PoC, cannot be transferred individually when considering a dynamic system response. Consequently, hybrid modeling is currently limited to modeling only simulations with steady state ending conditions. Harmonic power flow studies can require explicit conditions [63] such that e.g. the Kron reduction cannot always be applied to a nodal admittance matrix [64].
To develop hybrid model, the following options can be pursued:

- Option 1: Overcome the constraints of the Heisenberg-Gabor limit:

  Option 1a: By developing an iterative solver algorithm that can handle the physical divergent currents and voltages without running into convergence issues. This is rather a programming and mathematical issue and not scope of this work.

Option 1b: By clearly relating the solver convergence issue and the unsolvability of the equations to the physical signals, e.g. voltages and currents. They must not result from the implementation of the solver algorithm (bad solver parameters: e.g. too large step-size etc.).

- Option 2: Develop dynamic time-domain models:

  Option 2a: By describing the entire device in one model, e.g. a large-signal model that still represents the small-signal characteristics accurately. A promising approach can be the previously introduced use of ANNs.

  Option 2b: By transforming all small-signal frequency domain models into time-domain models and perform studies holistically by including all small-signal models.

Out of these options, only Option 2a has been tested by the author which leaves the other three options as further future work as well as a more detailed study of Option 2a.

Scalability of harmonic emission

For large-scale simulations and for network representations, the aggregation of inverters and the scalability of inverters is of interest to reduce the required computational power of the simulation and to ease the model representation.
It is annotated in [20] that multi-device studies have been performed in the past, e.g. [65, 66], whereby they are either bound to a small number of studied devices in time domain, or assume an overall linearity of the frequency model [61]. As the frequency model is only a small-signal model, the boundaries of the feasibility with regard to definitions of the small-signal characteristics are not disclosed. On the other hand, white-box models are not suitable for a large number of devices due to the increasing computational effort. Consequently, the author of this work has presented the scalability of frequency-domain models in [20] with the conclusion that the aggregation of inverters is feasible for small-signal studies, also in presence of resonances in the network impedance. For the aggregation, each inverter must be considered for the specifically studied operating power (that defines the operating point as explained in the later chapter 3) individually thus a single equivalent power model is not suitable due to the nonlinear dependency of the operating power on the voltage-current relation.

2.4.5 Harmonic stability analysis

This work demonstrates the dependency of the operating point on the small-signal model of the inverter. In literature, it is typically not disclosed for which operating point the harmonic stability analysis is performed (e.g. [61]) and further, what definition the author

applies for defining the reference point. In addition, the LV network is not appropriatly simplified, e.g. the representation of LV networks by an RL-equivalent is not suitable as it neglects resonances in the network impedance but often chosen. The impact of typical LV network characteristics, i.e. the impact of resonances [6] , is presented in the later of this work.
This work completes the interaction analyzes of single-phase inverters with classically represented, aggregated LV networks, i.e. the network impedance and the background voltage. The linearity study and the respective conclusions on the device aggregation at the PoC has to be done in future for all other types of grid-connected devices, e.g. rectifiers of houshold applications. This will give an understanding on the linearity of the network characteristics at individual PoCs and possible lead to new network representations. Especially with focus on future networks with a high penetration of PE devices, the linearity of public LV networks is assumed to change [13]. Follow-up studies on device aggregations will be another topic that has to be adressed in future.
Since even the resonance characteristics vary largely in public LV networks, the later of this work presents a probabilistic network consideration [1] in section 5.5.
With regard to commercially available devices, it is unrealistic to assume that all components can be represented in detail and with the exact parameters if solely based on data from the data sheet. Manufacturing deviations of physical elements and undisclosed control algorithms make an independent analyses impossible. Furthermore, the large number of manufacturer designs requires a more general analysis than the detailed study of one device with all its components as the number of available devices is too large to study each device by firstly implementing the detailed white-box model.
Consequently, the theoretic analysis of known designs is rather a theoretical but important part of the entire analysis of devices in the field. Since the device parameters are not disclosed or not reliable, the analysis of commercially available devices requires black-box based analysis and assessment methods. These methods rely rather on measurements based on the theoretic background knowledge. This background knowledge is developed in the following chapter 3 and its application on the measurements by optimizing the FS is explained in chapter 4.

Black-box simulations

For the black-box analysis, besides measurements and the formal analysis, simulations based on black-box models can possibly present a suitable approach. These models have to be dynamic and therefore in time domain though, dynamic black-box time-domain models have not been developed as of yet.

2.4.6 Assessment of overall stable operation

In practice, different scenarios of inverter shutdowns have been reported. Not all of these shutdowns relate to stability issues by classical means but rather other phenomena. A more general assessment of the inverter from a broader perspective is required that is related to the overall stable operation of the inverter including stability issues but also further impacts that trigger unwanted shutdowns. This will be addressed in section 5.6.

3 Linearity analysis

With regard to commercially available devices, the data generation for the stability assessment needs to be specified. While the state of the art provides a method in terms of the frequency sweep (section 2.2.3), the specific test signals have to be defined by taking the dependencies of the input signals on the small-signal characteristics of the inverter under consideration. Consequently, the internal signal processing has to be analyzed in more detail as an extension of the state of the art to provide a comprehensive background understanding for the later chapters, i.e. the measurement-based parametrization (chapter 4) and the stability analysis (chapter 5).
This chapter presents a qualitative analysis of the linearity of the individual, internal inverter components and consequently belongs to white-box analyses. As this is not applicable for commercially available inverters with unknown parameters, the focus of this chapter is on the general characteristics of the frequency spectrum of the grid-side current but does not include a quantitative analysis. Based on the gained knowledge, the representation of the small-signal characteristics of the inverter in terms of the LTP and the simplified LTI characteristics can be derived with regard to the input-output signal characteristics.

3.1 General signal processing

A general scheme of the input and output signals has been depicted previously in figure 2.1. It can be concluded that the control input signals are the measured, physical values $u_{\mathrm{PoC}}(t)$, $i_{\mathrm{PoC}}(t)$ and $u_{\mathrm{DC}}(t)$ and the control output signal is $u_{\mathrm{Inv\,ref}}(t)$ (see figure 2.3) that is finally converted by the modulator to the vector of the switching signals $\mathbf{sws}(t)$. The hardware input signals are $u_{\mathrm{PV}}(t)$, $i_{\mathrm{PV}}(t)$, $u_{\mathrm{PoC}}(t)$ and $\mathbf{sws}(t)$ that define $i_{\mathrm{PoC}}(t)$ as overall output signal. The interaction of the control with the hardware takes place at the inverter bridge. The bridge voltage is the convolution of $u_{\mathrm{DC}}(t)$ and $u_{\mathrm{Inv\,ref}}(t)$ that is transformed into $\mathbf{sws}(t)$ by the modulator.

3.2 Software components

As a general consideration on the control of commercially available single-phase inverters, it can be assumed that the individual components, e.g. the PLL, the DC-side control and the AC-side control, are based on linearizable controllers with respect to their small-signal characteristics, i.e. PI and PR controllers. The typical voltage distortions and the interaction of the inverter with the network impedance at the PoC by $i_{\text{PoC}}(t)$ cause harmonic power flows that are much smaller than the power related to the power frequency f_1. Thus the linearized small-signal model of the PI and PR controllers is suitable. All internal control parameters, i.e. the gains, are constant. In the last years, adaptive control algorithms and deadbeat controllers have been proposed in academic studies, e.g. [36], but not been found in low power commercially available inverters as of yet.
In practice, a required control deadtime is present compared to ideal switching and progression of the switching signal. This deadtime has a much smaller time constant than the studied effects in the harmonic range, e.g. at least a factor ten, so that the impact of the deadtime is not relevant for the harmonic frequency range.

Grid synchronization

Different methods to synchronize with the voltage at the PoC are possible such as assuming an ideal frequency of the LV network of 50 Hz (or 60 Hz respectively) and zero-crossing detection. Nowadays, single-phase inverters operate based on PLLs. In modern PLLs, a filter with linearizable small-signal characteristics adjusts the synchronization so that for a rather stable power frequency at the PoC, the PLL is analyzed by linear control theory. In practice, the bandwidth of the PLL has a major impact on the inverter impedance characteristics, e.g. in terms of passivity theory, as it defines the frequency range in which the control is active and when the PLL is too slow to track and adjust phase angle variations. For single-phase inverters, the PLL requires an orthogonal signal related to the input signal, i.e. the voltage at the PoC. This orthogonal signal has to be created artificially to enable the PLL to operate in $\alpha\beta$-frame or in dq-frame. Typically, the PLL of single-phase inverters operates in $\alpha\beta$-frame.

DC-link voltage control

The DC-link voltage control typically consists of a PI controller that ensures a stable DC-link voltage. Feed-forward and normalization are often applied as well as a 100 Hz Notch filter to suppress the 100 Hz voltage ripple for the underlaying control, i.e. the AC-current control. However, it is not necessary to implement any of these improvements though all three possibilities are typically not found together in one design.
A steep 100 Hz notch filter, i.e. with a narrow bandwidth, creates an unwanted sharp

slope in the inverter impedance characteristics though a smooth filter design will affect also the frequency region around 50 Hz which is unwanted. Anyhow, the 100 Hz Notch Filter cannot dampen the 2nd order harmonic ideally because its gain is not infinite, thus in practice, the 2nd order harmonics of DC-link voltage propagate partially to the AC side. Feed-forward will propagate dynamics of the DC-link voltage directly to the control output being typically added to the numerator while normalization is added to the denominator, thus having a reciprocal effect with regard to the propagation of a potential DC-link voltage distortion into the AC-current control.

AC-current control

The AC-current control can be assumed linear for the small-signal analysis based on its PI and PR controllers.

Modulator

The modulator itself will only cause an additional distortion in the supraharmonic frequency range above the switching frequency that result from its digital nature in real systems, i.e. being time-discrete and value-discrete.

Further software components

As mentioned in section 2, the auxiliary software algorithms run typically in the background and do not have an impact on the normal operation of the inverter.

Conclusion of software components

The software components including the control and respective software components are represented by linearized small-signal representations as current state of the art. With regard to a signal flow chart representation, the signal processing including respective controller feedback loops is handled by addition and subtraction.

3.3 Hardware components

Switches

The switches, i.e. the IGBTs, have nonlinear voltage-current characteristics and the switching frequency is independent of the voltage and current at the PoC. Nevertheless, with regard to the frequency range up to 2 kHz, the switching frequency is much larger for PV applications, e.g. a minimum of 5 kHz, and does not affect the small-signal models up to 2 kHz.

DC-DC converter

Typically, the DC-side voltage $u_{\mathrm{PV}}(t)$ is not high enough and is defined by the MPPT to enable optimal operating conditions on the DC-side panel strings. The DC-DC converter has a boost characteristic that sets $u_{\mathrm{DC\,av}}$ to a fixed value, typically in the range of 400 V to 600 V, as long as $u_{\mathrm{PV}}(t)$ is below the required voltage to enable a power flow from the DC side to the AC side, i.e. $u_{\mathrm{DC\,av}}$ is greater than $u_{\mathrm{PoC}}(t)$. The switching frequency of the DC-DC converter is typically very high, i.e. in the supraharmonic frequency range, and therefore out of the harmonic frequency range. Furthermore, the emission due to the switching has not been measured on the AC side. The decoupling of DC-side distortions and the AC-side current spectrum increases with an increasing frequency due to the DC-link capacitance. Therefore, the DC-side voltage $u_{\mathrm{PV}}(t)$ can be simplified for grid studies to only a DC component, i.e. 0 Hz.

DC-link capacitance

The DC-link capacitance is considered linear in the harmonic frequency range and resistive losses are neglected in the following.

AC-side filter circuit

The AC-side filter circuit is also considered with linear voltage-current characteristics in the harmonic frequency range. Current research studies analyze theoretically, if it is possible to implement coils with smaller inductances to save money and reduce the overall weight of the inverter. This causes the filter coils to saturate and challenges the linearity assumption for the overall AC-side filter circuit but no implementation in test stand inverters or commercially available inverters is known as of yet.

The filter circuit is typically designed as an LC-, an LCL-, or and $LCLC$-topology. All filter topologies have capacitors implemented to provide better filtering characteristics. The capacitance of the capacitors decouples high-frequent currents that are on one side generated by the inverter, e.g. at switching frequency, but also high-frequent currents from the LV network. Consequently, the high-frequent inverter impedance characteristics are typically passive.

Sometimes, additional filters to suppress supraharmonics, i.e. electromagnetic compatibility (EMC) filters are implemented but can be neglected in the harmonic frequency range.

Conclusion of hardware components

All hardware components can be assumed linear or negligible in terms of their voltage-current relation in the harmonic frequency range.

3.4 Modulation-based bridge voltage

Though both, the software and the hardware components can be linearized, their interaction in terms of $u_{\mathrm{Inv\,ref}}(t)$ or rather $\mathbf{sws}(t)$ and $u_{\mathrm{DC}}(t)$ will determine the modulation-based bridge voltage $u_{\mathrm{Inv}}(t)$. This interaction will be studied in detail in the following section to improve the understanding of the origin of frequency couplings in single-phase inverters.

3.4.1 DC-link voltage

Theoretic considerations

The DC-link voltage relates to the power flow and the buffered energy in the DC-link capacitance. The power flow at the DC-side at the connection to the PV modules can be formulated based on the scheme in figure 2.2 by

$$p_{\mathrm{PV}} = u_{\mathrm{PV}} i_{\mathrm{PV}} \tag{3.1}$$

assuming a constant power p_{PV} for the small-signal studies in the harmonic frequency range and neglecting voltage distortions on the PV strings. The power $p_{\mathrm{DC}}(t)$ at the DC link can be expressed by

$$p_{\mathrm{DC}}(t) = u_{\mathrm{DC}}(t) i_{\mathrm{DC}}(t) = p_{\mathrm{PV}} - p_{\mathrm{Cap}}(t), \tag{3.2}$$

for the power $p_{\mathrm{Cap}}(t)$ through the capacitance of the DC-link C_{DC}. It is important to notice, that the DC-link voltage $u_{\mathrm{DC}}(t)$ is not constant but can contain a significant ripple component as explained in the later of this chapter.

In (3.2), the losses in the switches and the copper losses are neglected. A comparison of power losses in three single-phase inverter topologies has been performed in [67] that has shown efficiencies better than 97.5 % for operating powers above 1 kW. Neglecting the inverter losses in the following, the power equilibrium demands the power $p_{\mathrm{DC}}(t)$ flowing into the inverter bridge and the power $p_{\mathrm{PoC}}(t)$ flowing into the LV network being equal with

$$p_{\mathrm{DC}}(t) = p_{\mathrm{PoC}}(t) \tag{3.3}$$

for

$$p_{\mathrm{PoC}}(t) = u_{\mathrm{PoC}}(t) i_{\mathrm{PoC}}(t) \tag{3.4}$$

and with sinusoidal AC-side voltages and currents

$$p_{\mathrm{PoC}}(t) = \hat{u}_{\mathrm{PoC}} \sin(2\pi f_1 t) \hat{i}_{\mathrm{PoC}} \sin(2\pi f_1 t) \tag{3.5}$$

or rather

$$p_{\mathrm{PoC}}(t) = \frac{\hat{u}_{\mathrm{PoC}}\hat{i}_{\mathrm{PoC}}}{2}(1 - \cos(4\pi f_1 t)) \tag{3.6}$$

and with (3.3)

$$p_{\mathrm{DC}}(t) = \frac{\hat{u}_{\mathrm{PoC}}\hat{i}_{\mathrm{PoC}}}{2} - \frac{\hat{u}_{\mathrm{PoC}}\hat{i}_{\mathrm{PoC}}}{2}\cos(4\pi f_1 t). \tag{3.7}$$

As mentioned in section 2.1.2, the phase angle $\varphi_{\mathrm{PoC}\,f1}$ between voltage and current at the PoC at power frequency has been measured with 0° thus the reactive power Q_{PoC} at the PoC is zero.
With a higher power generation on the DC side, $i_{\mathrm{PoC}}(t)$ increases and the power component of $p_{\mathrm{DC}}(t)$ that oscillates with $2f_1$ as annotated in (3.7) increases as well as the constant component of $p_{\mathrm{DC}}(t)$. This oscillating power has to be buffered in the DC-link capacitance so that the energy $e_{\mathrm{Cap}}(t)$ stored in the DC-link capacitance can be described with

$$e_{\mathrm{cap}}(t) = \frac{1}{2}C_{\mathrm{DC}}u_{\mathrm{DC}}(t)^2 \tag{3.8}$$

and

$$p_{\mathrm{Cap}}(t) = \frac{\mathrm{d}e_{\mathrm{Cap}}(t)}{\mathrm{d}t} \tag{3.9}$$

for

$$p_{\mathrm{Cap}}(t) = p_{\mathrm{PV}} - p_{\mathrm{DC}}(t) \tag{3.10}$$

and consequently

$$p_{\mathrm{Cap}}(t) = p_{\mathrm{PV}} - \frac{\hat{u}_{\mathrm{PoC}}\hat{i}_{\mathrm{PoC}}}{2} + \frac{\hat{u}_{\mathrm{PoC}}\hat{i}_{\mathrm{PoC}}}{2}\cos(4\pi f_1 t) = \frac{\hat{u}_{\mathrm{PoC}}\hat{i}_{\mathrm{PoC}}}{2}\cos(4\pi f_1 t). \tag{3.11}$$

For p_{PV}, $\hat{u}_{\mathrm{PoC}}$ and $\hat{i}_{\mathrm{PoC}}$ being constant, the maximum of the periodic change $\hat{p}_{\mathrm{Cap}}$ can be formulated with

$$\hat{p}_{\mathrm{Cap}} = \frac{\hat{u}_{\mathrm{PoC}}\hat{i}_{\mathrm{PoC}}}{2} = p_{\mathrm{PV}}. \tag{3.12}$$

According to (3.9), $p_{\mathrm{Cap}}(t)$ cannot have a constant part since the DC-link voltage would increase constantly until destroying the DC-link capacitor, so that the average power at the PoC $p_{\mathrm{PoC\,av}}$ has to equal the power of the PV modules p_{PV}, which is in accordance with the assumption that p_{PV} is constant for the considered time constants and

$$p_{\mathrm{Cap}}(t) = p_{\mathrm{PoC\,rip}}(t) \tag{3.13}$$

while

$$p_{\mathrm{PoC\,av}} = \max(p_{\mathrm{PoC\,rip}}(t)) \tag{3.14}$$

for steady state. In the following, $p_{\mathrm{PoC\,av}}$ will be also called operating power as it represents the power flow at the PoC averaged over a period of the power frequency.
The energy stored in the capacitance $e_{\mathrm{Cap}}(t)$ contains a constant component $e_{\mathrm{Cap\,const}}$ and a time-dependent component $e_{\mathrm{Cap\,rip}}(t)$, so that

$$e_{\mathrm{Cap}}(t) = e_{\mathrm{Cap\,const}} + e_{\mathrm{Cap\,rip}}(t) \tag{3.15}$$

and consequently

$$\frac{2}{C_{\mathrm{DC}}}(e_{\mathrm{Cap\,const}} + e_{\mathrm{Cap\,rip}}(t)) = u_{\mathrm{DC}}(t)^2 \tag{3.16}$$

with the arithmetic mean of the DC-link voltage $u_{\mathrm{DC\,av}}$ and

$$u_{\mathrm{DC}}(t) = u_{\mathrm{DC\,av}} + u_{\mathrm{DC\,rip}}(t). \tag{3.17}$$

The constant part $e_{\mathrm{Cap\,av}}$ must be proportional to the constant part of $u_{\mathrm{DC}}(t)^2$ so that

$$u_{\mathrm{DC}}(t)^2 = (u_{\mathrm{DC\,av}} + u_{\mathrm{DC\,rip}}(t))^2 = u^2_{\mathrm{DC\,av}} + 2u_{\mathrm{DC\,av}}u_{\mathrm{DC\,rip}}(t) + u^2_{\mathrm{DC\,rip}}(t). \tag{3.18}$$

For a normal operation of the inverter, $u_{\mathrm{DC\,rip}}$ is much smaller than $u_{\mathrm{DC\,av}}$ and $u^2_{\mathrm{DC\,rip}}(t)$ can be neglected. Therefore,

$$\frac{2}{C_{\mathrm{DC}}}e_{\mathrm{Cap\,rip}}(t) \approx 2u_{\mathrm{DC\,av}}u_{\mathrm{DC\,rip}}(t) \tag{3.19}$$

so that the time series $u_{\mathrm{DC\,rip}}(t)$ and $e_{\mathrm{Cap}}(t)$ consist of the same frequency spectra. Considering (3.9), $p_{\mathrm{Cap}}(t)$ consists also of the same frequency spectrum according to the sum rule for derivatives.
The previous findings demonstrate the demand to reformulate (2.1) to

$$\begin{aligned} u_{\mathrm{Inv}}(t) =& \frac{4}{\pi}\sum_{i=0}^{\infty}\hat{u}_{\mathrm{DC}\,i}\cos(2\pi i f_1 t)\sum_{n=1}^{\infty}\frac{J_n\left(n\frac{f_1}{f_{\mathrm{car}}}\frac{\pi}{2}m_{\mathrm{a}}\right)}{n\frac{f_1}{f_{\mathrm{car}}}}\sin\left(n\frac{\pi}{2}\right)\cos(2\pi f_1 t)+ \\ & \frac{4}{\pi}\sum_{i=0}^{\infty}\hat{u}_{\mathrm{DC}\,i}\cos(2\pi i f_1 t)\sum_{m=1}^{\infty}\sum_{k=-\infty}^{\infty}\frac{1}{q}J_{2k-1}\left(q\frac{\pi}{2}m_{\mathrm{a}}\right)\cos([m+k-1]\pi)\cdot \\ & \cos(4m\pi f_{\mathrm{car}}t + [2k-1]2\pi f_1 t). \end{aligned} \tag{3.20}$$

To conclude the previous findings, the analytic formulation describes that even for an undistorted voltage at the AC side, the DC-link voltage $u_{\mathrm{DC}}(t)$ is not constant but has frequency components at frequencies above 0 Hz as described in (3.20). In detail, three impact factors can be identified that affect the frequency spectrum of $u_{\mathrm{Inv}}(t)$ mainly:

1. The power of the PV modules p_{PV} since it affects the distortion $u_{\mathrm{DC\,rip}}(t)$ in $u_{\mathrm{DC}}(t)$.

2. The DC-link capacitance C_{DC} which is a fixed internal parameter of the inverter.

3. The modulation index m_{a} that represents the relation of $u_{\mathrm{DC\,av}}$ and the fundamental frequency component $\underline{U}_{\mathrm{PoC}}(f_1)$.

The first two findings have been used to identify the DC-link capacitance for future grey-box modeling by a noninvasive measurement of $u_{\mathrm{DC\,rip}}(t)$ and p_{PV} as described in section D.2.

Measuremental validation

For a measuremental validation, it is a challenge to open the inverter case in practice and further to gain access to the clamps on both sides of the inverter bridge, e.g. to measure $u_{\mathrm{DC}}(t)$ or $u_{\mathrm{Inv}}(t)$. Though for inverter I, exemplary measurements have been made, e.g. for the identification of the AC-side filter circuit (see appendix A.1.1), inverter II and inverter III could not be measured with open cases. For one, because there were safety mechanisms implemented that prevent the inverter operation with open case. Furthermore for inverter II, the clamps inside where covered by further components of the inverter.
For a qualitative validation of the previous theory with regard to the impact of the DC-link voltage and the power p_{PV} from the PV panels, i.e. the operating power, the grid-side current has been measured. This grid-side current is excited by the voltage at the inverter bridge and thus contains similar frequency components as the voltage at the inverter bridge.
For these measurements, an undistorted (sinusoidal) voltage has been applied at the PoC at different operating powers, i.e. from 500 W to the individually rated power of the inverters, i.e. 3 kW for inverter III and 4.5 kW for inverter I and inverter II. However, inverter I required at least 1 kW to operate so that for this inverter, no data was measured at 500 W. No additional network impedance was applied.
The measured response at the PoC is depicted in Figure 3.1.
It can be seen that for a sinusoidal voltage at the PoC, there are frequency components at multiples of the power frequency, i.e. harmonics, in the current at the PoC. The current amplitudes decrease for the power frequency and also for the harmonics for inverter II for a decreasing operating power, i.e. bright color of the circles indicate a low operating power and dark circles a high operating power. The amplitudes of the currents have been plotted until 1 kHz as they decrease in general with an increasing frequency if the voltage at the PoC is sinusoidal at power frequency. In contradiction, inverter I shows an alternating dependency on the operating power between the amplitudes of the odd and even order current harmonics. Though the design of the inverters is unknown, this indicates the differences between the implementations and possibly different manufacturer

approaches on handling the dynamics of the DC-link voltage (see reformulation of $u_{\mathrm{Inv}}(t)$ in (3.20)).

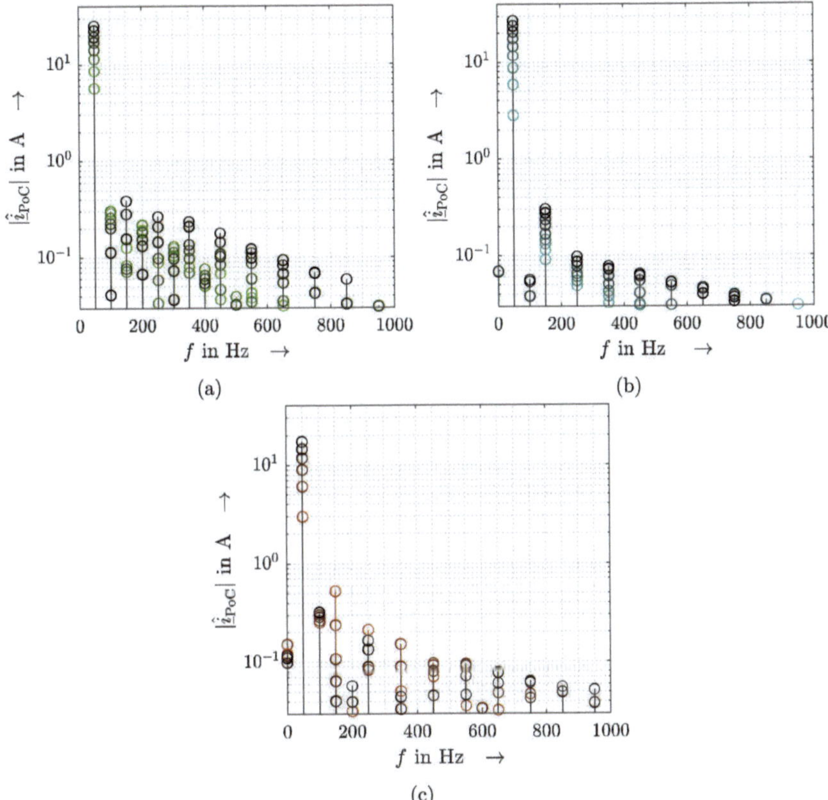

Figure 3.1: Impact of p_{PV} on $i_{\mathrm{PoC}}(t)$ from low power (bright color) in 500 W steps to rated power (dark color) of inverter I from 1 kW to 4.6 kW in (a), of inverter II from 500 W to 4.6 kW in (b) and inverter III from 500 W to 3.0 kW in (c).

3.4.2 AC-side voltage

The AC-side voltage is physically applied at the AC-side terminals and also serves as an input signal for the control (see Figure 2.3), e.g. for the PLL and the AC-current control. Noninvasive black-box identification methods on commercially available devices have not been developed for the control algorithms as of yet though $u_{\mathrm{Inv\,ref}}(t)$ represents the control output. According to figure 2.3, $u_{\mathrm{Inv\,ref}}(t)$ will be converted to **sws** to trigger the switches while $u_{\mathrm{Inv\,ref}}(t)$ contains the line-neutral voltage that is intended by the control.

Higher frequencies

In practice, it is unrealistic to assume an undistorted supply voltage, i.e. a sinusoidal waveform of $u_{\mathrm{PoC}}(t)$. Due to other grid-connected devices with a nonlinear voltage-current relation, $u_{\mathrm{PoC}}(t)$ contains frequency components next to f_1. This distortion contains typically frequency components of integer multiples of the power frequency, i.e. harmonics at order h if the power frequency is also the fundamental frequency. However, also frequency components that are not integer multiples of f_1 can be present in the frequency spectrum. These frequency components can result e.g. from deadtimes of the control, from the switching of the IGBTs or simply from the emission of other grid-connected devices.
As previously explained, a distortion in $u_{\mathrm{PoC}}(t)$ will propagate through the control. It is possible to mitigate this distortion in the frequency region where the control is active which is consequently limited to the bandwidth of the PLL, i.e. some hundred Hertz. In practice, a perfect cancellation is not realistic. Therefore, a distortion in $u_{\mathrm{PoC}}(t)$ affects $u_{\mathrm{Inv\,ref}}(t)$ so that assuming a purely sinusoidal reference signal as in (2.1) is not realistic in the field. The analytic description of $u_{\mathrm{Inv}}(t)$ has to be further adapted regarding the modulation parameter q which is actually frequency dependent and must be adapted such that the frequency-dependent modulation parameter q_h considers any harmonic frequency $f_{\mathrm{ref}\,h}$ of the reference signal of the voltage at the inverter bridge and the carrier frequency of the pulse width modulation of the IGBTs. Consequently,

$$q_h = 2m + [2n-1]\frac{f_{\mathrm{ref}\,h}}{f_{\mathrm{car}}}, \quad m,n \in \mathbb{N} \tag{3.21}$$

can be formulated and thus

$$\begin{aligned} u_{\mathrm{Inv}}(t) =& \frac{4}{\pi}\sum_{h=1}^{\infty}\sum_{i=0}^{\infty} \hat{u}_{\mathrm{DC}\,i}\cos(2\pi i f_{\mathrm{DC}\,i}t + \varphi_{\mathrm{DC}\,i})\cdot \\ & \left(\sum_{n=1}^{\infty} \frac{J_n\left(n\frac{f_{\mathrm{ref}\,h}}{f_{\mathrm{car}}}\frac{\pi}{2}\frac{\hat{u}_{\mathrm{ref}\,h}}{\hat{u}_{\mathrm{DC}\,0}}\right)}{n\frac{f_{\mathrm{ref}\,h}}{f_{\mathrm{car}}}} \sin\left(n\frac{\pi}{2}\right)\cos(2n\pi f_{\mathrm{ref}\,h}t + n\varphi_{\mathrm{ref}\,h}) + \right. \\ & \sum_{m=1}^{\infty}\sum_{k=-\infty}^{\infty} \frac{1}{q_h} J_{2k-1}\left(q_h\frac{\pi}{2}\frac{\hat{u}_{\mathrm{ref}\,h}}{\hat{u}_{\mathrm{DC}\,0}}\right)\cos([m+k-1]\pi)\cdot \\ & \left. \cos(4mk\pi f_{\mathrm{car}}t + [2k-1](2\pi f_{\mathrm{ref}\,h}t + \varphi_{\mathrm{ref}\,h}))\right) \end{aligned} \tag{3.22}$$

so that m_{a} is expanded to the ratio of all real-value amplitudes $\hat{u}_{\mathrm{ref}\,h}$ of the reference signal to the amplitude of the DC component $\hat{u}_{\mathrm{DC}\,0}$ of the DC-link voltage. Furthermore in (3.22), f_1 is expanded to all frequency components in the frequency spectrum of $u_{\mathrm{Inv\,ref}}(t)$. In (3.21), only harmonics are considered for q_h though in practice, interharmonics are typically negligible in LV networks. For an analytic description of the frequency spectrum of (3.22), the complex amplitudes can be calculated by making use of the convolution operator $*$ with

$$\begin{aligned}\underline{\hat{u}}_{\mathrm{Inv}}(f_{\mathrm{Inv}}, h, i, k, m, n) = &\frac{4}{\pi}\underline{\hat{u}}_{\mathrm{DC}\,i}(f_{\mathrm{DC}\,i}) * \\ &\left(\underline{\hat{u}}_{\mathrm{LF}}(n f_{\mathrm{ref}\,h}) + \underline{\hat{u}}_{\mathrm{HF}}(2mk f_{\mathrm{car}} + [2k-1] f_{\mathrm{ref}\,h})\right)\end{aligned} \tag{3.23}$$

for the complex amplitudes of the frequency spectrum of $u_{\mathrm{DC}}(t)$ in terms of $\underline{\hat{u}}_{\mathrm{DC}\,i}(f_{\mathrm{DC}\,i})$ and for the complex amplitudes in the low frequency (LF) region, i.e. the band around the fundamental frequency, by means of

$$\underline{\hat{u}}_{\mathrm{LF}}(n f_{\mathrm{ref}\,h}) = \frac{J_n\left(n \frac{f_{\mathrm{ref}\,h}}{f_{\mathrm{car}}} \frac{\pi}{2} \frac{\hat{u}_{\mathrm{ref}\,h}}{u_{\mathrm{DC}\,0}}\right)}{n \frac{f_{\mathrm{ref}\,h}}{f_{\mathrm{car}}}} \sin\left(n \frac{\pi}{2} e^{\mathrm{j} n \varphi_{\mathrm{ref}\,h}}\right) \tag{3.24}$$

and the complex amplitudes of the HF region, i.e. the so-called frequency bands at entire multiples of the carrier frequency in terms of

$$\underline{\hat{u}}_{\mathrm{HF}}(2m f_{\mathrm{car}} + [2k-1] f_{\mathrm{ref}\,h}) = \frac{1}{q_h} J_{2k-1}\left(q_h \frac{\pi}{2} \frac{\hat{u}_{\mathrm{ref}\,h}}{\hat{u}_{\mathrm{DC}\,0}}\right) \cos\left([m+k-1]\pi\right) e^{\mathrm{j}(2k-1)\varphi_{\mathrm{ref}\,h}}. \tag{3.25}$$

For a detailed analysis, specifically in the HF-range, the analysis in time domain is very time-consuming and challenges the computational power of standard office computers. The main reason is the large number of nested sum-terms over a broad frequency range. The calculation of the spectrum in frequency domain according to (3.23) reduces the required computational power and time for the calculation significantly which enables detailed studies in the supraharmonic frequency range.

Impact on supraharmonics

As a consequence of (3.22), a distortion in $u_{\mathrm{Inv\,ref}}(t)$ will not only be present in the LF range but also mirror into the supraharmonic frequency range, which is defined in this work as above 2 kHz up to 150 kHz. In present standards and literature, e.g. [68], this coupling is not considered and the harmonic and the supraharmonic frequency range are typically assessed separately.

This frequency coupling into the supraharmonic frequency range is measurable and has been published by the author in [12]. A frequency sweep is performed in the laboratory at the AC-side inverter clamps of inverter II in 10 Hz steps from 60 Hz to 2 kHz at a phase angle $\varphi_{\mathrm{PoC}\,U}$ of the sweep component of $0°$ in relation to the component at 50 Hz for an amplitude of $2\sqrt{2}$ V being superimposed to a sinusoidal voltage with an amplitude of 325 V at 50 Hz.

Figure 3.2 depicts the frequency spectra of the applied voltage at the PoC for each measurement indicated by the n° of the measurement. As examples, for the first measurement where n° is one, only the frequency of 50 Hz is excited (reference point), i.e. as described with an amplitude of 325 V. For the second measurement where n° is two, the frequency of 50 Hz is excited with an amplitude of 325 V and an additional 60 Hz component, i.e. the sweep component, is excited with $2\sqrt{2}$ V.

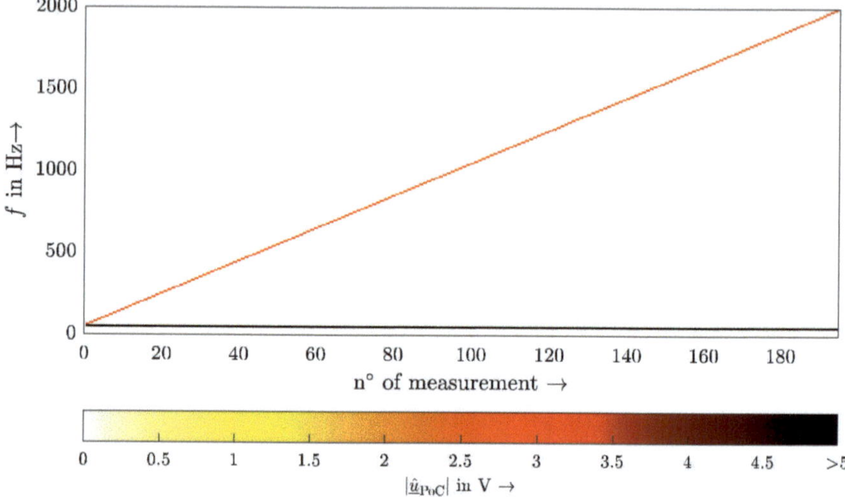

Figure 3.2: Measurement-specific frequency spectra of the applied amplitudes of the voltage at the PoC [12].

The propagation through the control and the effect on the bridge voltage up to the first frequency band around the effective switching frequency at 16 kHz are depicted in figure 3.3 (a).
However, as previously mentioned, harmonic instabilities have only been reported in the frequency range where the control is active, i.e. below 2 kHz.

3.5 Inverter representation

For black-box studies, the current components and their origin can be separated into dependent and independent current components. If the current components are independent of the external input signals, e.g. the emission due to PWM-based switching at a fixed frequency, the representation as a voltage source can be chosen [68]. These independent current components are not part of this work and rather present in the supraharmonic frequency range. Dependent current components can be represented either by admittances

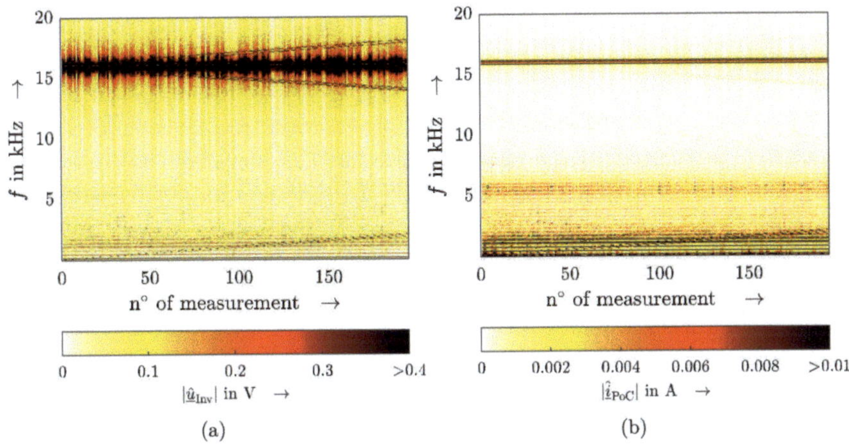

Figure 3.3: Measurement-specific frequency spectra of the voltage amplitudes at the inverter bridge (a) and the current amplitudes at the PoC (b) [12].

and impedances (LTI characteristics), or by controlled current sources derived from the FCMs (LTP characteristics).

3.5.1 Overview of impact factors

From a black-box perspective, the impact factors can be related to the input signals on both sides of the inverter, i.e. the DC side and the AC side. It is possible to relate an individual input signal to multiple impact factors, e.g. the AC-side voltage can be specified in frequency domain according to its frequencies, amplitudes and the phase angles of the frequency components in relation to the power frequency. An overview of the dependency of each impact factor on the time-periodic characteristics of the inverter is listed in table 3.1.

Table 3.1: Overview of impact factors.

Side	**Input signal**	**Impact factor**	**Dependency**
DC side	Operating power	Power p_{PV}	Nonlinear
AC side	Voltage at the PoC	Frequency f_U	Nonlinear
		Amplitude $\hat{u}_{\mathrm{PoC}}$	Linear
		Phase angle $\varphi_{\mathrm{PoC}\,U}$	Linear

The impact factors are identified for a two-pole representation of the inverter connected to an electric network by the AC-side terminals, i.e. at the PoC. Assuming all internal parameters of the inverter to remain constant, i.e. the design of the inverter does not

change, the external signals can be identified as the operating power and the voltage at the AC-side.
The established theory regarding linearizable software and hardware together with the analytic description developed in the previous section implies the frequency-dependent linearity of the voltage-current relation. This classifies for a simplified small-signal representation in frequency domain as LTI model by averaging over the power frequency or by neglecting the frequency coupling terms as classical admittance and for the more advanced representation as LTP model by including the frequency coupling terms by FCMs. On the other hand, the operating power affects the current at the PoC even if the background voltage does not change. The operating power is considered quasi-stationary due to the speed of clouds and the moving of their shadows on the ground. Changes in the operating power resulting from the DC side are much slower, i.e. at least some hundred milliseconds. Thus the time constants are typically much larger and relate to subharmonics compared to time constants in harmonic studies. The quasi-stationary consideration of the operating power is bound to the application. Applications other than PV-based power generation can require dynamic operating power considerations if the change of the operating power affects the harmonic frequency range.
With regard to table 3.1, the dependency of the current at the PoC on the operating power has been visualized in figure 3.1 for the laboratory measurements. The dependency on the voltage frequency become obvious for the LTI and LTP results in the following section 3.5.4. While the author did not present measurement results of the linear relation between the amplitude $\hat{u}_{\mathrm{PoC}}$ and the phase angle $\varphi_{\mathrm{PoC}\,U}$ of the voltage and the respective currents at the PoC, the measurements presented in the later chapter 4 in section 4.3.2 demonstrate the applicability of the principle of superposition, thus the linearity. Further, as mentioned throughout the different studies, different amplitudes have been used for the test voltages but concluded on the same small-signal inverter model.

3.5.2 Available devices

The available simulation models are based on a modular model and labeled by capital Latin letters, e.g. A, B etc. For more details see section A.1. For the laboratory measurements, the three commercially available inverters are labeled in Roman numbers from I to III. The measurement results and the assessment of these three commercially available inverters are presented in the following of this chapter and have been published with regard to the impact of the operating power in [17].
In addition, the exemplary LTI characteristics, i.e. the inverter impedances, of four simulated inverter models created by the modular modeling approach are shown in figure A.6 in appendix A.1.5.

3.5.3 Measurement specifications

The measurement data has been generated at the AC-side inverter clamps for a sinusoidal reference voltage with an amplitude of 325 V at 50 Hz while the inverter response, i.e. the current at the PoC has been measured. The parameters of the voltage frequency sweep component have been set with an amplitude of $\sqrt{2}{\cdot}5\,\text{V}$ and a phase angle $\varphi_{\text{PoC}\,U}$ of $0\,°$ while the frequency sweep is performed in 50 Hz steps. Measurements for other parameters and reference voltages have been performed, e.g. as previously introduced with only $\sqrt{2}\cdot 2\,\text{V}$, but are not further considered as they all implied the validity of the analytic study, i.e. a nonlinear dependency on the operating power but a linear dependency regarding the AC-side voltage amplitudes and phase angles of the sweep component, and consequently have yield to the LTI and LTP representations. In practice for single-phase PV inverters, the amplitude of $\sqrt{2}\cdot 5\,\text{V}$ is a suitable amplitude for a single-frequent sweep component and recommended by the author. Too small voltage excitations increase the impact of the measurement uncertainties, i.e. see appendix B.2.4 while too large amplitudes can cause the inverter to shut down [17], i.e. as related to chapter 5.6.
The frequency sweep has been performed for different operating powers of the inverters, i.e. from 500 W (inverter I requires more than 500 W to operate and was therefore not measured at 500 W) in 500 W steps to the rated power of the inverters which is for inverter I and inverter II 4.6 kW and for inverter III 3.0 kW and the currents and voltages at the PoC have been measured to calculate the LTI and the LTP characteristics.

3.5.4 Measurement results

The presentation of the measurement results can be sorted according to their LTI characteristics and their LTP characteristics.

Linear time-invariant characteristics

The LTI characteristics, e.g. as shown by admittances in figure 3.4 and impedances in figure 3.5 relate to the LTP charactersistics, e.g. represented by an FCM, such that

$$\underline{Z}(f) = \frac{1}{\underline{Y}_{\text{Inv}}(f,f)}, \forall f_I = f_U = f. \tag{3.26}$$

Consequently, the LTI characteristics are represented by the main-diagonal elements of the FCM.
The admittances in figure 3.4 and the impedances in figure 3.5 represent the LTI small signal characteristics of the inverters for the measured operating powers. For the impedance-based stability analysis, it is rather the impedance than the admittance that is considered

though the relation between the admittance and the impedance is well known by

$$\underline{Y}_{\text{Inv}}(f) = \frac{1}{\underline{Z}_{\text{Inv}}(f)} \tag{3.27}$$

so that the depicted impedances of the inverters in figure 3.5 are calculated based on the same measurements as the admittances in figure 3.4.

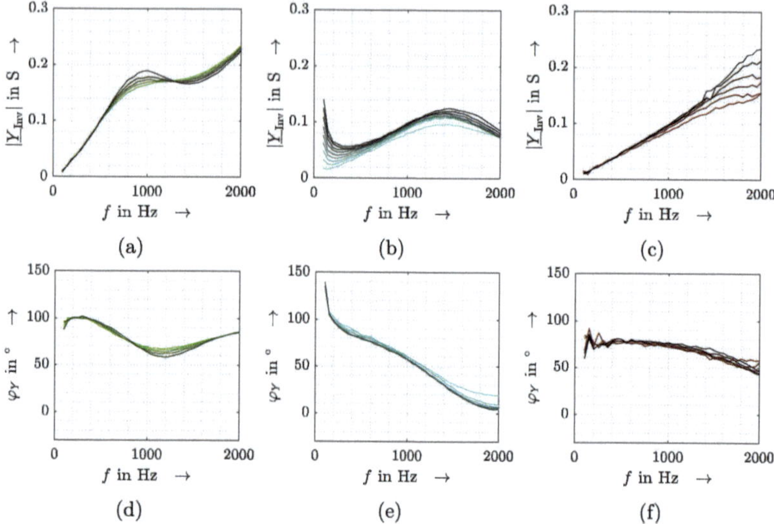

Figure 3.4: Magnitude $|\underline{Y}_{\text{Inv}}|$ (a-c) and phase angle φ_Y (d-f) of inverter admittances from low power (bright color) in 500 W steps to rated power (dark color) of inverter I from 1 kW to 4.6 kW in (a, d), of inverter II from 500 W to 4.6 kW in (b, e) and inverter III from 500 W to 3.0 kW in (c, f) based on [17].

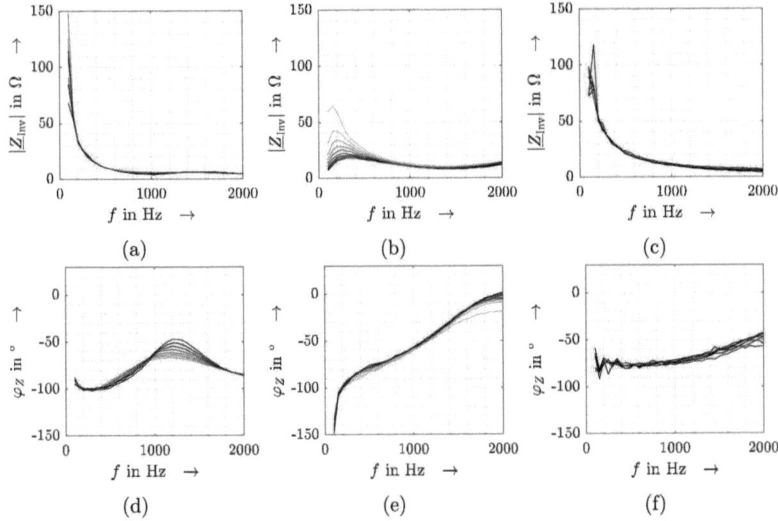

Figure 3.5: Magnitude $|\underline{Z}_{\text{Inv}}|$ (a-c) and phase angle φ_Z (d-f) of inverter impedances from low power (bright color) in 500 W steps to rated power (dark color) of inverter I from 1 kW to 4.6 kW in (a, d), of inverter II from 500 W to 4.6 kW in (b, e) and inverter III from 500 W to 3.0 kW in (c, f) based on [16].

Linear time-periodic characteristics

The LTP characteristics in frequency domain, represented by the operating-point dependent FCMs are depicted in figure 3.6.

Compared to the LTI representation, the LTP representation by the heatmap enables to visualize the high sensitivity of inverter III for all measurements in the frequency range around 1400 Hz which is furthermore increasing with an increasing operating power.
The main-diagonal elements of the FCMs of single-phase PV inverters are typically significantly larger than the off-diagonal elements such that the main-diagonal elements are simply annotated as above the range of interest [19], e.g. in figure 3.6. The amplitude of the frequency coupling components, i.e. the amplitudes of the elements of the off diagonals decrease the further they are apart from the main diagonal. With regard to the frequency couplings at multiples of ±100 Hz , only the ±100 Hz frequency couplings are significantly present. Already for ±200 Hz, the amplitude values are mostly negligible.

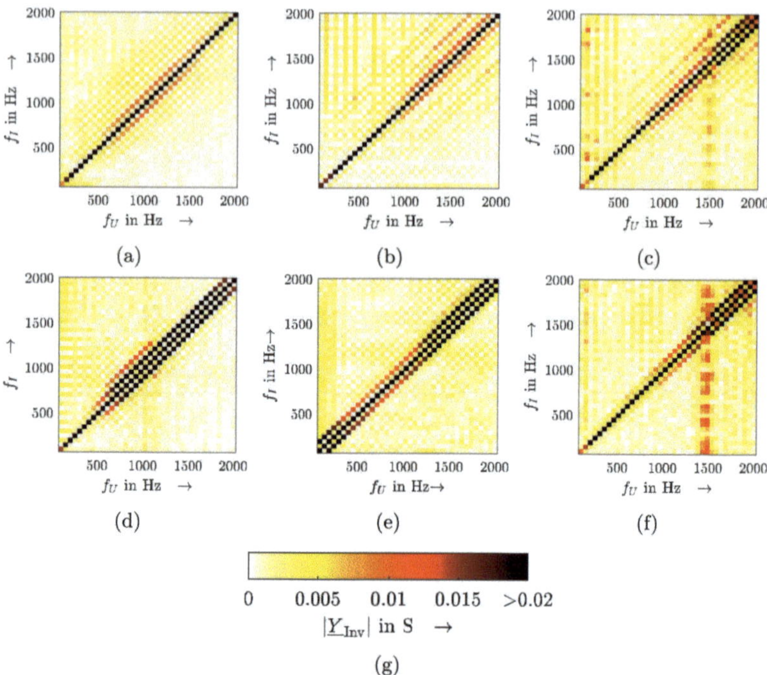

Figure 3.6: Frequency coupling matrix of inverter I with rated power of 4.6 kW for 1 kW(a) and 4.5 kW (d), of inverter II with rated power of 4.6 kW for 0.5 kW (b) and 4.5 kW (e), for inverter III for 0.5 kW (c) and 3 kW (f) with rated power of 3.0 kW and the respective scale (g) [17].

3.5.5 Measurement assessment

The measurement results at the different operating powers of the inverters can also be assessed by their LTI characteristics and their LTP characteristics individually as presented by the author in [17].

Linear time-invariant characteristics

To evaluate the impact of the operating power, the maximum absolute deviation $\mathrm{devabs}_{|\underline{Y}|}$ of the magnitudes can be calculated with

$$\mathrm{devabs}_{|\underline{Y}|}(f) = \max_{P}\left(|\underline{Y}_{\mathrm{Inv}}|(f, P)\right) - \min_{P}\left(|\underline{Y}_{\mathrm{Inv}}|(f, P)\right), \tag{3.28}$$

the maximum absolute deviation of the phase angles $\text{devabs}_{\varphi Y}$ with

$$\text{devabs}_{\varphi Y}(f) = \max_{P}(\varphi_Y(f,P)) - \min_{P}(\varphi_Y(f,P)), \tag{3.29}$$

and the maximum relative deviation $\text{devrel}_{|\underline{Y}|}$ of the magnitudes with

$$\text{devrel}_{|\underline{Y}|}(f) = \frac{\max_{P}(|\underline{Y}_{\text{Inv}}|(f,P)) - \min_{P}(|\underline{Y}_{\text{Inv}}|(f,P))}{\max_{P}(|\underline{Y}_{\text{Inv}}|(f,P))}. \tag{3.30}$$

Since the rated values at rated power are not necessarily the minimum, or maximum values, the relative deviations have always been related to the maximum values and not the values at rated power, which accounts for 3.30 but also for 3.32 with regard to the assessment of the LTP characteristics.

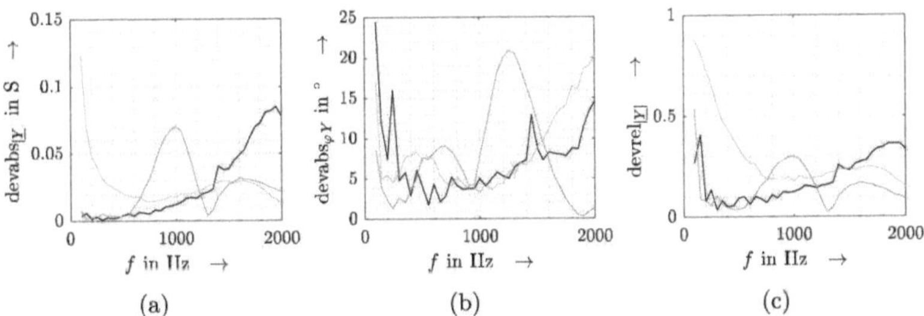

Figure 3.7: Maximum absolute deviations $\text{devabs}_{|Y|}$ of LTI characteristics of inverter I (green), inverter II (cyan) and inverter III (chestnut) based on [17].

Significant differences of the inverter-specific maximum deviations of the admittance magnitudes can be observed. Inverter II indicates a high maximum absolute deviation of 123 mS at 100 Hz and a maximum relative deviation of 88 % at 100 Hz that decreases with an increasing frequency. The maximum relative deviations of inverter I and inverter III are rather small, e.g. the maximum relative deviation of inverter I stays below 50 %, but with regard to the entire frequency range up to 2 kHz more pronounced in the range of the resonances. [17]

Linear time-periodic characteristics

To assess the LTP characteristics of the different FCMs, the maximum absolute deviation between all operating powers can be calculated elementwise by

$$\text{devabs}_{|\underline{Y}|}(f_I, f_U) = |\max_{P}(\underline{Y}_{\text{Inv}}(f_I, f_U, P)) - \min_{P}(\underline{Y}_{\text{Inv}}(f_I, f_U, P))| \tag{3.31}$$

as depicted in figure 3.8. The maximum absolute deviation of the main-diagonal elements of the FCM is typically much bigger than the off-diagonal elements, i.e. the frequency coupling components. One reason is simply their larger absolute element values.

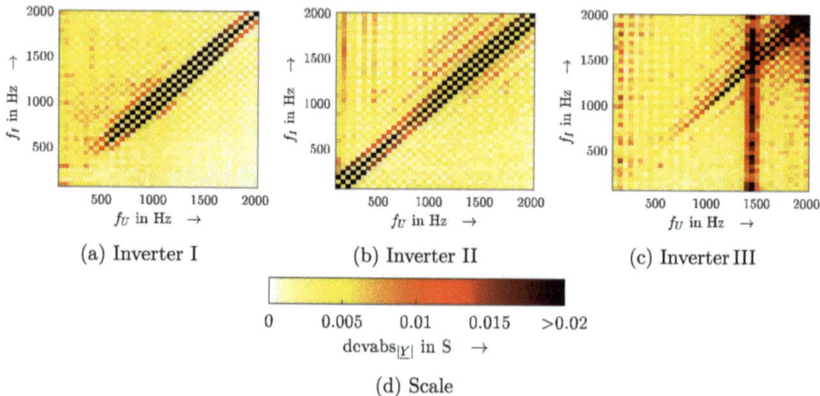

(a) Inverter I (b) Inverter II (c) Inverter III

(d) Scale

Figure 3.8: Maximum absolute deviations devabs$_{|\underline{Y}|}$ of LTP characteristics and respective scale [17].

Also, the maximum relative deviation can be calculated by relating the minimum value of each element of the FCM to the maximum value of this element by

$$\mathrm{devrel}_{|\underline{Y}|}(f_I, f_U) = \frac{\max_P\left(|\underline{Y}_{\mathrm{Inv}}(f_I, f_U, P)\right) - \min_P\left(\underline{Y}_{\mathrm{Inv}}(f_I, f_U, P)|\right)}{|\max_P\left(\underline{Y}_{\mathrm{Inv}}(f_I, f_U, P)\right)|} \tag{3.32}$$

and is visualized in figure 3.9. A threshold of 10 mA is set for the evaluation of the maximum relative deviation due to the impact of the measurement uncertainty (cf. section B.2.4). The maximum relative deviation of the off-diagonal elements at ±100 Hz, is however much stronger than the maximum relative deviation of the main diagonal elements. As presented in the beginning of this chapter, the power buffering of the DC-link capacitance and therefore the operating power has a dominant impact on the DC-link ripple voltage. The DC-link voltage of inverter II has been measured and is exemplarily depicted in figure D.9. Based on the measurement results can be assumed that inverter I and inverter III can suppress the propagation of the DC-link ripple voltage into the grid-side currents since the sidebands in the LF range, e.g. up to 600 Hz for inverter I and approximately 850 Hz for inverter III, are negligible even for higher operating powers. In practice, this becomes possible with a 100 Hz notch filter in the control though this work did not develop an identification method for specific control implementations so that the assumption if the 100 Hz notch filter cannot be proven.

More general, the implementation of the control in its various designs affects the overall

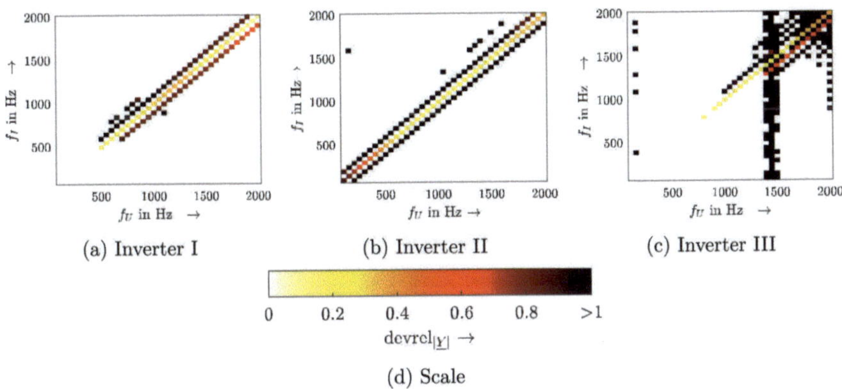

(a) Inverter I

(b) Inverter II

(c) Inverter III

(d) Scale

Figure 3.9: Maximum relative deviations $\mathrm{devrel}_{|\underline{Y}|}$ of LTP characteristics and respective scale [17].

impact of the operating power on the small-signal characteristics. Due to the cascaded control structure and a large variety of possible implementations, no non-invasive approach has been presented to identify the unknown topology of the control and the respective parameters for commercially available (black-box) inverters.

It becomes also visible, that inverter III has an unusually high sensitivity around 1.5 kHz which is not comparable to the other two measured inverters or the simulated inverters. More precisely, the maximum relative tolerance for all frequency coupling components at 1.5 kHz is above 100 %. However, as can be seen in figure 3.6, the respective off-diagonal elements of the FCM are still below 15 mS while the main-diagonal elements (LTI characteristics in figure 3.4) indicate values at 1.5 kHz above 120 mS.

3.5.6 Power cuboid

The current visualization of the FCMs is the heatmap as previously depicted. The FCM implies a 2-dimensional tensor, i.e. a matrix. Based on the findings of this chapter, it is rather appropriate to extend the 2-dimensional representation by one dimension. The resulting tensor of 3rd order can be visualized by a cube or rather a cuboid depending on the order of the represented voltages, currents and operating powers [17]. The discrete aggregation of the FCMs in the cuboid as represented can possibly be used for interpolation to create a more refined lookup table if necessary. Alternatively, the general analytic description of the element-specific dependency on the operating power seems also suitable in future.

It remains a challenge to show the entire data, e.g. the individual FCMs at different operating powers, simultaneoulsy within one power cube, but in practice, the cube can be shifted and tilted and the layers can be spread further apart to get an overview of the data.

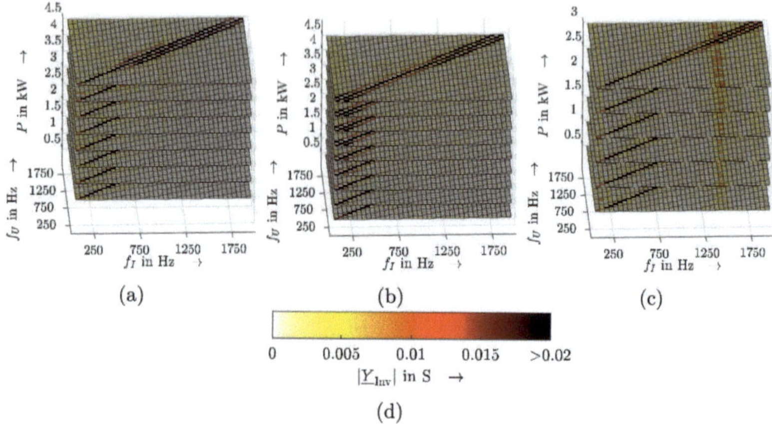

Figure 3.10: Power cuboid of inveter I with rated power of 4.6 kW (a), inverter II with rated power of 4.6 kW (b), inverter III with 3.0 kW (c) and respective scale (d) [17].

3.5.7 Relevance of operating power for harmonic power flow studies

The dependency of the voltage-current characteristics of the inverter on the operating power affects the results of harmonic power flow studies. As depicted in figure 3.7, figure 3.8 and figure 3.9, the deviation between different operating powers is inverter-dependent and even further for each inverter frequency-dependent. The generation of detailed measurement data, i.e. at different operating powers, should therefore be taken into account for the measurement-based identification.

3.6 Conclusions on the linearity analysis

The main conclusions of the linearity analysis will be used in the following two chapters and are therefore specifically summarized to emphasize the main findings:

The operating power has a nonlinear impact on the LTI and the LTP representation of the inverters by means of the voltage-current relation at the PoC and thus for a holistic representation, the inverter should be measured at different operating powers, e.g. in 500 W steps up to the rated power.

Different amplitudes for the single-frequent sweep component of the frequency-sweep-based identification method have been tested leading to the same LTI and LTP represen-

tations. An amplitude of $\sqrt{2} \cdot 5\,\mathrm{V}$ has been identified as suitable for all inverters and is recommended by the author.
The voltage at the PoC has a nonlinear frequency-dependent impact on the current at the PoC but a linear impact regarding the amplitudes of the voltage distortion components and the respective phase angles $\varphi_{\mathrm{PoC}\,U}$.
Consequently, the currently established FCM as a 2D representation of the small signal characteristics of an inverter can be extended to a 3D representation, e.g. a power cuboid.
In detail, the analytic description of the state of the art has been extended by expanding the DC-link voltage to a realistic representation in real devices. This has been achieved by deriving the impact of the operating power on the DC-link voltage, but also by including the distortion of the voltage at thePoC and the power flow at the PoC into the analytics. Consequently, the new description is based on a frequency-dependent modulation paramter q_h and the expandion of the modulation index m_{a} to all frequencies of the distortion of the DC-link voltage.
This explains the presence of frequency coupling components qualitatively thus indicates that these frequency components are also linear regarding the amplitude and phase angle of the voltage distortion components in the voltage at the PoC but nonlinearily dependent on the frequency of these components.
The operating power has a nonlinear impact on the frequency coupling components.
The analytical description leads to the conclusion that the disortion in the harmonic frequency range affects also the supraharmonic frequency range of the voltage at the inverter bridge though in practice, the AC-side filter circuit suppresses the propagation into the LV network.
The maximum absolute values of the frequency coupling components by means of the off-diagonal elements of the FCM, i.e. the LTP characteristics, are much smaller than the main-diagonal elements of the FCM, i.e. the LTI characteristics. The maximum relative deviation between one off-diagonal element at different operating powers is much bigger than the maximum relative deviation of one main-diagonal element at different operating points.

4 Measurement-based parametrization

4.1 General considerations

As introduced previously, the LTI description represents the averaged LTP small-signal model but can also be derived by neglecting the periodic changes. However, the superimposed current frequency components resulting from the frequency couplings have to be considered during the identification, so that a unity step, e.g. often used for the identification of linearizable systems, cannot be applied as identification method. A second reason is that due to the small current amplitudes in the current response, a strong voltage excitation would be required to excite the entire current frequency range sufficiently, i.e. accurately measurable current amplitudes. This would challenge the operation of the inverter causing the inverter to possibly shut down.

4.2 Frequency-sweep-based parametrization

As revised in the state of the art in chapter 2, the principle of the FS has been introduced in general in literature. However, as there are individual characteristics with regard to dependencies on different sets of input signals that relate to the applications for which the PE devices are used, the generally introduced FS needs to be specified for parameterizing an individual type of PE devices, e.g. single-phase inverters for PV applications. Chapter 3 has disclosed that the applied voltage at the PoC has a frequency-dependent impact on the small-signal current response but can be linearized in terms of its amplitudes and phase angles.
As a consequence of chapter 3, (e.g. see table 3.1), it is sufficient to measure the different operating powers, e.g. in 500 W steps, only for one voltage amplitude and one phase angle at each applied frequency, e.g. the harmonic frequencies in 50 Hz steps up to 2 kHz. The author's experience can be summarized by stating that too small voltage amplitudes for the sweep component in the applied signal increase the impact of the measurement uncertainty while the maximum voltage amplitude is limited by the immunity and the individual sensitivity of the inverters. Specific inverters have shut down for harmonic voltage amplitudes of 10 V during laboratory measurements. As concluded already from chapter 3, an amplitude of $\sqrt{2} \cdot 5\,\mathrm{V}$ has been found a good trade-off for which a phase

angle of 0 ° is chosen though the choice of the phase angle does not have any restriction. More general based on the author's experience, for single-frequent sweep components, this value of $\sqrt{2} \cdot 5\,\mathrm{V}$ can be used also for other grid-connected devices that are operated in public LV networks in Europe besides single-phase PV inverters. The author has explicitly tested rectifiers such as implemented in electric vehicle chargers, battery storage systems, lighting equipment and power supply for computers. The author has performed test measurements where devices have shut down when the amplitude of the single-frequent distortion of the voltage at the PoC was 13 V.

4.3 Multi-frequent parametrization using interharmonics

A challenge of the identification method via the FS is the long measurement duration caused by the single-frequent sweep component. Each individual measurement point has to be set by the voltage generator and a steady state has to be reached before the recording of that measurement point can start. Afterwards, each measurement point has to be stored.

As a simple example from the authors considerations in [11]: For a set of FCMs that shall represent the device characteristics in the harmonic range, i.e. for a maximum number $n_{h\,\mathrm{FS}}$ of 40 voltage harmonics and the number $n_{p\,\mathrm{FS}}$ of the tested different operating powers being 5 (20 % steps with regard to the rated power), this sums up to 200 measurement points. Assuming a time duration t_{FS} of 5 s per measurement point (loading the signal vectors, applying the voltage, setting the steady state, including data storage, etc.) adds up to 16.6 min. If a dependency of the amplitude and phase angle of the AC-side voltage of the inverter on the FCM has to be taken into account, even only for validation purposes, the number of required measurements increases drastically as the new impact factor factorizes the overall measurement points. For the number $n_{\varphi\,\mathrm{FS}}$ of tested phase angles being 12 and the number of tested amplitudes of the sweep component $n_{\hat{u}_{\mathrm{FS}}}$ being five, the number N_{FS} of required measurements can be calculated with

$$N_{\mathrm{FS}} = n_{h\,\mathrm{FS}} n_{p\,\mathrm{FS}} n_{\varphi\,\mathrm{FS}} n_{\hat{u}_{\mathrm{FS}}} \tag{4.1}$$

to 1200 measurements and results by multiplication with t_{FS} to an overall measurement duration of 60,000 s, i.e. 16 h and 40 min, for one device. The importance for a fast approach (FA) becomes obvious, especially, if an operating point of a device can only be maintained for a limited duration. In the following section, such an FA is presented and published by the author. [11]

4.3.1 Rule-based definition of test voltage frequencies

With regard to the small-signal characteristics, the origin that dominates the frequency couplings in single-phase inverters has been found in the convolution of the DC-link voltage and $u_{\mathrm{Inv\,ref}}(t)$ as demonstrated in chapter 3. Compared to other effects that lead to nonlinear frequency couplings, e.g. rectifying and frequency couplings resulting from switching, this convolution leads to a linear frequency coupling over the entire frequency range, e.g. up to 2 kHz. Frequency couplings are also present for interharmonic excitations while their current responses are dependent on the power flow, i.e. the dominating power frequencies. A generalization of the applicable voltage frequencies to the entire frequency range, i.e. not only harmonics, enables the use of interharmonics for the identification of the inverter impedance characteristics. One of the main advancements of the FA is the multi-frequent distortion compared to the single-frequent distortion in the voltage. The frequency bin size and the excited voltage frequencies need to be selected carefully to ensure that the resulting current frequency components do not fall into the same frequency bin. Therefore, rules are defined so that the resulting current frequency components can be related to the respective exciting voltage frequency components.

Rule 1

The voltage frequencies of one test measurement must be apart by a large number of multiples of f_1 to prevent that the frequency coupling components in the current that result at $\pm 2nf_1$ are overlapping. The expected number of dominant frequency coupling components in the current is typically below three, i.e $n < 3$, cf. figure 3.6. The amplitude of the frequency coupling components in the current decreases with an increasing n as explained and demonstrated in chapter 3, so that Rule 1 can be formulated with

$$n \geq 3, n \in \mathbb{N} \tag{4.2}$$

which has been identified to define the minimum frequency spacing between the voltage frequencies in one test measurement. [11]

Rule 2

The sums and differences of two test frequencies of the voltage, i.e. of any frequency tuple of the vector $\boldsymbol{f_U}$, should not result in f_1 [11]. The beating components at frequencies of f_1 and the components at f_1 itself can become indistinct. Analytically speaking, the application of the modulu operation by $\boldsymbol{f_U} \bmod f_1$ must give different results for the entity of all tuples of the excited frequencies. and can be formulated by

$$(f_2 \pm f_3) \bmod f_1 \neq f_1, \quad f_2, f_3 \in \boldsymbol{f_U}. \tag{4.3}$$

Rule 3

The sums and differences of any j-tuple of $\boldsymbol{f_U}$ should be different as often as possible to avoid similar beating frequencies. To reduce the number of required measurement points and because of significantly lower energy in the beating frequencies aside from f_1, the number of j-tuples with the same sum or difference should not exceed two. [11]

Rule 4

For obtaining the minimum number of measurement points, the entire frequency range of interest, i.e. up to 2 kHz, should be used for each measurement point for the set of excited frequencies.

4.3.2 Application

The FA is tested on inverter I for 50 % of its rated power, i.e. 2.3 kW. For the sampling rate of 1 MHz and a measurement interval of 1 s, the frequency resolution is 1 Hz. Under optimal conditions, 49 frequencies can be applied at the same time according to rule 2. Due to a frequency uncertainty of the programmable power amplifier, e.g. see appendix B, a frequency spacing of 5 Hz is chosen between applied voltage frequencies resulting in a similar difference between the expected frequency coupling components in the current [11]. Therefore, the maximum number of frequency components in $\boldsymbol{f_U}$ is reduced in practice compared to the theoretically possible number.

Test signals

The applied voltage frequencies, i.e. the power frequency and the superimposed test frequencies, chosen for the measurement points are listed in table 4.1.

Table 4.1: Tested frequencies based on [11].

Measurement	Power frequency in Hz	Test frequencies in Hz
M1	50	215; 480; 645; 855; 1020; 1285; 1400; 1535; 1665; 1830; 1995
M2	50	165; 430; 595; 805; 970; 1325; 1350; 1485; 1615; 1780; 1945

The entire identification consists of two measurement points (measurement M1 and M2) each containing 12 voltage frequency components. Instead of 39 identified harmonics, only 22 frequencies are applid. As shown in the later, the interpolation over these components still leads to sufficiently accurate results though more frequency components could be

applied as long as the defined rules are fulfilled. A larger number of tested applied frequencies, i.e. a larger number of measurements and a larger overall measurement duration, provides more intermediate values for the later interpolation. [11]
For superimposed frequency components of an amplitude of $\sqrt{2} \cdot 5\,\text{V}$, the THD is too high so that the inverters are not able to operate stable and shut down. Based on the author's experience, an amplitude of $\sqrt{2} \cdot 2\,\text{V}$ has been identified as suitable for each tested frequency. A phase angle of the tested frequency component is set with a phase angle $\varphi_{\text{PoC}\,U}$ of $0\,°$ in relation to the component at power frequency which is applied with an amplitude of $\sqrt{2} \cdot 230\,\text{V}$. [11]
As mentioned previously, the reduction of the amplitude of the sweep components leads to stronger measurement deviations, e.g. see the measurement results in the later section 4.3.2. However, if the measurement accuracy is too small, the number of test frequencies has to be reduced that are applied at the same time, or the measurements have to be repeated and averaging methods have to be applied. Both, the reduction of simultaneously applied test frequencies and the repetition of measurements lead to a longer overall measurement duration.

Measurement evaluation

The measurement evaluation is performed in six steps:

1. Determine excited frequencies
   The excited voltage frequencies have to be determined. This can either be achieved by using the vector of the reference signal, or by analyzing the measurement data. Using the real measurement data instead of the vector of the reference signal, small frequency deviations between the output of the programmable power amplifier and the intended reference signal, are also included. With regard to the measurement data, the excited voltage frequency components can be determined automatically by setting a threshold based on the chosen amplitude, e.g. a threshold of $2\,\text{V}$ for an excitation of $\sqrt{2} \cdot 2\,\text{V}$.

2. Identify corresponding frequency coupling components
   The corresponding current frequency components can be calculated with

$$f_{I+}(f_U, n) = f_\text{U} + 2nf_1, \quad n \in \mathbb{N} \tag{4.4a}$$

$$f_{I-}(f_U, n) = f_\text{U} - 2nf_1 \tag{4.4b}$$

   and consequently

$$f_I = \{f_{I+}, f_{I-}\}. \tag{4.5}$$

   A respective visualization is depicted in figure 4.1.

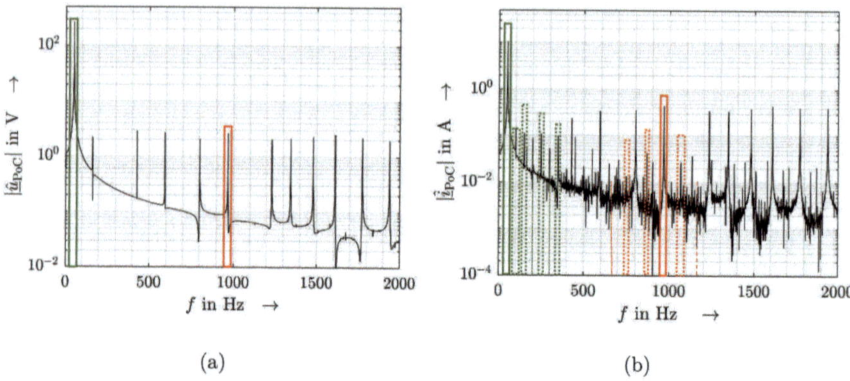

(a) (b)

Figure 4.1: Amplitudes of voltage (a) and current (b) of measurement M2 with impact of the voltage component at power frequency (green in (a)) on the respective currents (green in (b) at the same frequency (solid line) and other frequencies (dashed lines)) and the highlighted voltage distortion component at 970 Hz (red in (a)) on the respective currents (red in (b) at the same frequency (solid line) and other frequencies (dashed lines)) based on [11].

3. Calculate $\underline{Y}_{\text{Inv}}(f_I, f_U)$
   The values of $\boldsymbol{Y}_{\text{Inv}}$ can be calculated elementwise in terms of

$$\underline{Y}_{\text{Inv}}(f_I, f_U) = \frac{\underline{I}(f_I)}{\underline{U}(f_U)} \tag{4.6}$$

   for all excited voltages at frequencies f_U and all related currents at the respective current frequencies f_I. Compared to the FS, no superposition with regard to a reference point has to be considered, since according to the proposed rules, it has been avoided to have superimposed current frequency components [11]. Each current frequency can be related explicitly to one voltage frequency, e.g. the red marked current components in figure 4.1 (b) relate only to the voltage component at 970 Hz marked in red in 4.1 (a) and the green marked current components in figure 4.1 (b) relate only to the voltage component at power frequency, i.e. 50 Hz, marked in green in 4.1 (a).

4. Create vectors for respective n
   The excited voltage frequencies will each lead to multi-frequent current responses, e.g. marked as individual sets of colored crosses („x“) aligned along the thin black (vertical) lines, i.e. in columns in figure 4.2. The LTI characteristics are represented for n being zero, i.e. the main diagonal of the scheme in figure 4.2. Shifting from

the main diagonal up or down leads to the off-diagonal elements, i.e. the LTP characteristics with n being nonzero. All calculated elements of (4.6) with the same n can be grouped together.

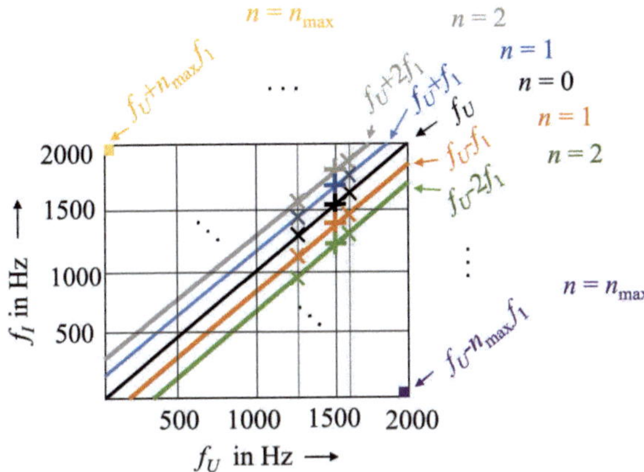

Figure 4.2: Scheme of the interpolation based on [11].

5. Interpolate along the diagonals and create admittance characteristics
   Compared to the FS, the excited frequencies of the FA are in most cases not the harmonics. To recalculate the data for all harmonics, the grouped values related to the same n-value can now be interpolated, i.e. along the colored diagonals in figure 4.2. The interpolation between two measured values $Y(f_{I+0}, n)$ and $Y(f_{I+1}, n)$ at the respective frequencies f_{I+0} and f_{I+1} is performed linearly for a value at f_{I+} with

$$Y(f_{I+}, n) = Y(f_{I+0}, n) + \frac{Y(f_{I+1}, n) - Y(f_{I+0}, n)}{f_{I+1} - f_{I+0}} f_{I+} \tag{4.7}$$

   and for a value at f_{I-} with

$$Y(f_{I-}, n) = Y(f_{I-0}, n) + \frac{Y(f_{I-1}, n) - Y(f_{I-0}, n)}{f_{I-1} - f_{I-0}} f_{I-}. \tag{4.8}$$

   The admittance characteristics at the specific harmonic orders are represented by the interpolated values of the main diagonal, i.e. n is zero. It is also possible to determine the interpolated values for the diagonals with n being nonzero, which would enable to identify the off-diagonal elements of the respective FCM. As the magnitude of the off-diagonal elements is significantly smaller than the main diagonal elements, their calculation uncertainty is assumed large. [11]

A quantitative analysis and improvements on the identification of the off-diagonals is left to future work.

Measurement results

The measurement results of the FA are depicted in figure 4.3 together with the results from the FS. The similarity between the results of both approaches is visible though

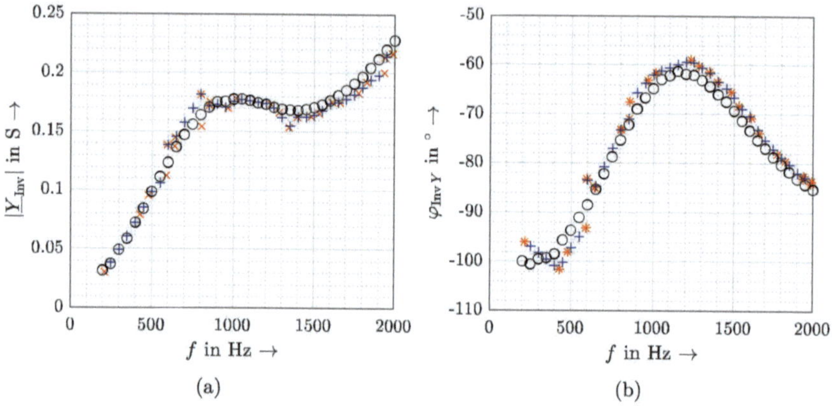

Figure 4.3: Admittance Magnitude $|\underline{Y}_{\mathrm{Inv}}|$(a) and phase angle $\varphi_{\mathrm{Inv}\,Y}$ (b) characteristics of inverter I parametrized by the FS („o“, black) and the FA in terms of the performed measurements(„x“, red) and the interpolated measurements („+“, blue) [11].

the deviations of the magnitude $\mathrm{err}_{\mathrm{Inv}\,|\underline{Y}|}$ between the admittance magnitudes $|\underline{Y}_{\mathrm{Inv\,FS}}|$ parametrized by the FS and the admittance magnitudes $|\underline{Y}_{\mathrm{Inv\,FA}}|$ parametrized by the FA are calculated with

$$\mathrm{err}_{\mathrm{Inv}\,|\underline{Y}|}(f) = \frac{|\underline{Y}_{\mathrm{Inv\,FA}}(f)| - |\underline{Y}_{\mathrm{Inv\,FS}}(f)|}{|\underline{Y}_{\mathrm{Inv\,FS}}(f)|} \tag{4.9}$$

and the deviations of the phase angles $\mathrm{err}_{\varphi\,\mathrm{Inv}\,Y}$ between the phase angle of the inverter admittance $\varphi_{\mathrm{Inv}\,Y\mathrm{FS}}$ parametrized by the FS and the phase angle $\varphi_{\mathrm{Inv}\,Y\,\mathrm{FA}}$ parametrized by the FA with

$$\mathrm{err}_{\varphi\,\mathrm{Inv}\,Y}(f) = \varphi_{\mathrm{Inv}\,Y\,\mathrm{FS}}(f) - \varphi_{\mathrm{Inv}\,Y\,\mathrm{FA}}(f). \tag{4.10}$$

As shown in figure 4.4, the maximum magnitude deviation is 11.7 % and the maximum phase angle deviation is 5 ° thus indicating reasonable results assuming reliable results performed by the FS that is used as reference. The maximum magnitude deviation of 11.7 % is at the frequency of 600 Hz. From 600 Hz to 800 Hz, this deviation is higher than for the other frequencies. In this range, only one test frequency, i.e. 645 Hz, is applied in the test voltages (compared to five, i.e. from 600 Hz to 800 Hz for the normal frequency sweep). Consequently, the linear interpolation requires more measured points for a higher accuracy in this frequency range at the cost of a longer overall measurement duration, e.g. by a third measurement. On the other hand, the dependency of the operating power shows, that a different operating power can affect the admittance characteristics by more than 12 %, e.g. see figure 3.7, thus for a first application of the new method, the results are acceptable.

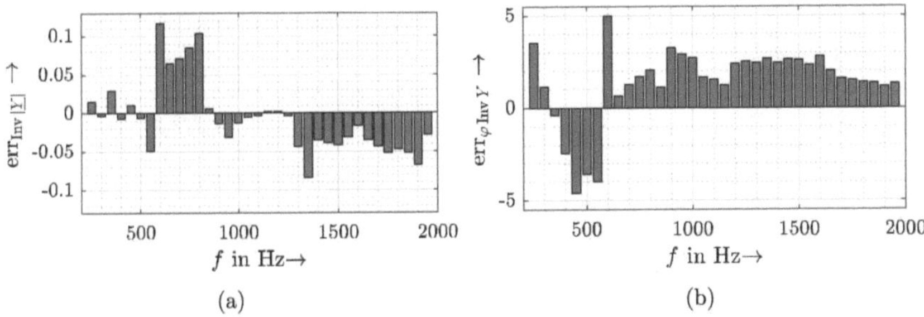

Figure 4.4: Deviations of admittance magnitude $\text{err}_{\text{Inv}\,|\underline{Y}|}$ (a) and phase angle $\text{err}_{\varphi\,\text{Inv}\,Y}$ (b) of inverter I based on [11].

4.3.3 Further considerations

If the system is linear, e.g. with regard to the voltage at the PoC, it is possible to identify it even in case of superimposed frequencies as analytically described in appendix E which enables field identification approaches.

For these measurements in the field, it is often a challenge to have the same measurement conditions over a long test period, e.g. switching operations within the LV network during the measurement can affect the measurement data and the voltage at the PoC can change not only based on the injected current by the inverter but also when the background voltage of the LV network changes. However, the LTI and the LTP characteristics can be identified even for a changing voltage at the PoC as they are linear depending on this voltage.

In chapter 3, the dependency of single-phase PV inverters on the quasi-stationary operating power, i.e. the solar irradiance, has been shown. Non-invasive field measurements become possible in future as explained in section E.2.3 and section E.2.4 in case the operating power is also measured or derived from current and votlage at power frequency and the results are assigned later to the respective operating powers to create the small signal models for the individual operating powers.

4.4 Conclusions on the measurement-based parametrization

By including the results from the linearity analysis in chapter 3, the overall measurement duration for the frequency sweep can be reduced significantly as it is sufficient to apply one amplitude and one phase angle for the sweep component in the voltage at the PoC at each frequency step for the different operating powers.
Since the impact of the voltage distortion at the PoC is linear, the principle of superposition can be used to improve the measurement-based parametrization further by applying multiple frequencies in the voltage distortion simultaneously. To achieve this, rules have to be followed in case the resulting current frequencies shall be assigned to only one voltage frequency. The amplitude of the exciting frequency components in the voltage distortion have to be reduced so that the THD at the PoC is not too high, e.g. below 8 % according to EN 50160 [38], which could cause the inverters to shut down.
It is even possible that different voltage components lead to a current response at the same frequency, e.g. if the rules presented in this chapter are not applied. Theoretically, it is possible to assign both current components at this one frequency in reverse to the voltages as shown in appendix E but might cause challenges in practice du to small current responses for frequency coupling components.

5 Harmonic black-box stability analysis

The harmonic stability of inverters will be analyzed in the following based on the knowledge gained in the previous chapters.

5.1 Formal analysis

As presented in section 2.3.3, the harmonic stability analysis is typically related to the Eigenvalue analysis or the impedance-based analysis. The Eigenvalue analysis is bound to knowledge about the zeros and poles of the transfer function characteristics of input to output signals. However, the measurement-based analysis only provides discrete values for the small-signal characteristics (the FCMs and impedances or rather admittances), but no continuous analytic description. Fitting algorithms that identify the poles and zeros have not been applied or proposed as of yet. The success of an applied fitting algorithm would be directly related to correctly assuming a suitable order of the transfer function characteristics. In practice, the order of the transfer function characteristics, i.e. the order of the small-signal model, is related to the inverter design (hardware and software components). Consequently, the correct model order varies between different inverters.

5.1.1 Impedance-based analysis

However, the impedance-based stability analysis that is based on the Nyquist stability allows to make use of the discrete measurement-based values and will be applied in the following. The general transfer function of the inverter, a two-pole representation at the PoC, can be described with

$$\underline{I}_{\mathrm{PoC}}(f) = \frac{\underline{I}_{\mathrm{PoC\,ref}}(f)\underline{Z}_{\mathrm{Inv}}(f)}{\underline{Z}_{\mathrm{Inv}}(f) + \underline{Z}_{\mathrm{g}}(f)} - \frac{\underline{U}_{\mathrm{g}}(f)}{\underline{Z}_{\mathrm{Inv}}(f) + \underline{Z}_{\mathrm{g}}(f)}, \tag{5.1}$$

which can be derived from figure 2.4 by considering the reciprocal relation of the inverter admittance and impedance (e.g. as formulated in 3.26) and is consequently based on the LTI model, e.g. as disclosed by the author in [19]. As mentioned previously, the time-invariant model can be interpreted as the averaged time-periodic model for the periodicity with regard to power frequency and its entire multiples.

All time-domain simulations and all measurements at the test stand inherently include this LTP characteristics. If the results of the theoretic LTI analysis, e.g. in frequency domain, and the validation by time-domain simulations and physically measured inverters conclude the same findings, it is appropriate to neglect the time-periodic characteristic in the analsyis method. In addition, the frequency coupling components (LTP characterisitcs) of single-phase inverters for PV applications are typically below 10 % compared to the steady state averaged (LTI characteristics) components at the same frequencies.

With regard to the LV network impedance, no measurement of the LTP characteristics has been performed. Specific reference measurements have proven the presence of the periodic changes but have not been quantified [69]. For the underlying background theory, the analysis that has been performed in chapter 3 is required for all grid-connected devices and the network equipment. Specifically rectifiers are known to be dependent on the voltage waveform at the PoC. Their small-signal characteristics in frequency domain are highly nonlinear and have an impact on the linearity of the overall network impedance. The device aggregation of different sets and types of devices, e.g. combinations of rectifiers, inverters and converters, as well as a suitable representation of the LV network will have to be studied in future work. Since there is no data regarding the periodic changes of the LV network impedance available, the LV network is represented in the following by its classical representation based on the state of the art as introduced in chapter 2 with a passive network impedance and the background voltage.

The background voltage is considered independent of the injected power of the inverter. Consequently, the background voltage distortion presents a disturbance, e.g. in the open loop transfer function, but its contribution is constant.

To analyze the system stability, the impedances of both, the LV network and the inverter, are of relevance. To point out the relevance of the impedances, (5.1) can be reformulated to

$$\underline{I}_{\mathrm{PoC}}(f) = \left(\underline{I}_{\mathrm{PoC\,ref}}(f) - \frac{\underline{U}_{\mathrm{g}}(f)}{\underline{Z}_{\mathrm{Inv}}(f)}\right) \frac{1}{1 + \frac{\underline{Z}_{\mathrm{g}}(f)}{\underline{Z}_{\mathrm{Inv}}(f)}}. \tag{5.2}$$

Assuming a reasonable initial current at the PoC and a noticeable inverter impedance $\underline{Z}_{\mathrm{Inv}}$, the current spectrum at the PoC is affected by the complex ratio of the network impedance $\underline{Z}_{\mathrm{g}}$ and the inverter impedance $\underline{Z}_{\mathrm{Inv}}$ [19]. If this complexe ratio is -1, the current at the PoC becomes very large and leads to an instability of the inverter in the respective frequency region. The magnitude of 1 of this ratio results when both impedances have the same magnitude. The negative sign requires an active frequency region in the inverter impedance, since by definition, the network impedance is passive over the entire frequency range. Nowadays, first LV networks show active frequency regions, e.g. for a high penetration by PE devices. However, these networks are not considered in this work and have to be analyzed in more detail.

5.1.2 Passivity theory

Similar to the previous analysis, passivity relates to the same analysis approach. Passivity theory states that if the closed-loop transfer function of a system is passive, the entire system is stable. According to the currently applied state of the art, the LV network is considered passive, which can be expressed with regard to its phase angle $\varphi_{\mathrm{g}\,Z}$ by

$$|\varphi_{\mathrm{g}\,Z}(f)| < 90^{\circ} \tag{5.3}$$

and

$$Re\{\underline{Z}_{\mathrm{g}\,Z}(f)\} > 0. \tag{5.4}$$

As mentioned in the previous chapters, the bandwidth of the control is smaller than 1 kHz and defines the active frequency region with

$$|\varphi_{\mathrm{Inv}\,Z}(f)| > 90^{\circ} \tag{5.5}$$

or rather

$$Re\{\underline{Z}_{\mathrm{Inv}}(f)\} < 0 \tag{5.6}$$

of the inverter impedance characteristics. Field measurements have shown a first parallel resonance in the impedance characteristics of public LV networks below 1 kHz [70]. According to passivity theory, the phase margin φ_{PM} is essential for the stability and can be calculated by

$$\varphi_{\mathrm{PM}} = 180^{\circ} - \varphi_{\mathrm{g}\,Z} - \varphi_{\mathrm{Inv}\,Z} \tag{5.7}$$

indicating a theoretically stable system, if

$$\varphi_{\mathrm{PM}} > 0. \tag{5.8}$$

Due to the phase angle shift from the active frequency region to the passive capacitive frequency region of the inverter impedance, highly inductive network impedances are most critical for the harmonic stability of the inverter, whereby for highly inductive network impedances, the magnitude increases with an increasing frequency [10]. On the other hand, the network impedance magnitudes are small compared to the inverter impedance magnitudes. An intersection of the inverter impedance magnitudes and the network impedance magnitudes is most likely in highly inductive networks. Higher resistive parts increase the network impedance magnitude but dampen the voltage and current oscillations which dampens the overall system so that for networks with significant resistive characteristics, the phase margin criterion is typically not violated. Measurement impedances have been derived from measurements in a measurement campaign in [70]

and are disclosed in the appendix in figure C.2.
The critical frequency region can be assumed below the first resonance of the network impedance [1]. The absolute value of the phase angle of the inverter decreases in that frequency region while the magnitude of the network impedance increases. For further analyses, the frequency range below the first resonance of the network impedance is of interest and is approximated by an RL-equivalent, e.g. [10]. The following approach is based on this finding.

5.2 Simplified measurement-based analysis

The formal analysis requires knowledge of the inverter impedance characteristics. To identify the impedance characteristics of commercially available inverters, the test stand requires the advanced voltage generator that has been used for the measurements presented in chapter 3 and chapter 4 to apply the specific voltage distortion accurately. In practice, such equipment is often not available and implies high test stand costs. It is of practical interest to develop a method without requiring an accurate voltage generator. Therefore, a method is presented in this section as an alternative measurement-based approach to assess the harmonic stability of an unknown inverter and published in [15] as an extension of [10]. This novel approach accounts also as a validation for the applicability of the previously described formal analysis (see section 5.1) on commercially available single-phase inverters, though the presented approach based on the formal analysis as presented in the previous section has also been disclosed by the author in [14].
The previous measurements have demonstrated a generally decreasing inverter impedance magnitude for all inverters in the frequency range up to 1 kHz (e.g. figure 3.5). Varying the test stand impedance in the laboratory that represents the network impedance in the field, the inverter operation is challenged stronger for a larger test stand inductance L_{test}, i.e. a highly inductive network impedance. The systematic increase of L_{test} leads to an amplified current response of the inverter in the critical frequency region. The critical frequency region of the interaction of the inverter and the test stand impedance can be identified by the dominating current frequency components.

5.2.1 Test stand setup

The test stand consists of an inverter that is connected on the DC side to a DC power generator and on the AC side to a voltage generator that provides the background voltage, i.e. the programmable power amplifier (see appendix B). In addition, a test stand impedance $\underline{Z}_{\text{test}}(f)$ is now installed between the voltage generator and the inverter with a voltage $\underline{U}_{Z\,\text{test}}$ and surge protection devices (SPDs) are installed on both sides of the test stand impedance. The test stand setup is shown in figure 5.1.

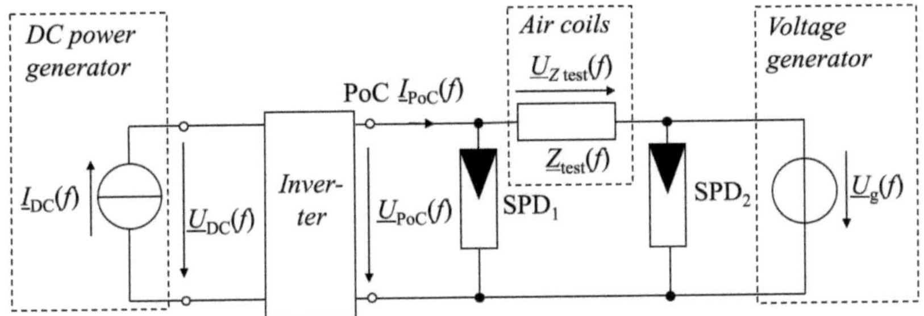

Figure 5.1: Test stand setup for a simplified measurement-based stability analysis based on [10].

The overall test stand impedance can be calculated based on the sum of the test inductances L_{test} with the test inductances L_1 and L_2 and the respective magnetic coupling M_{mag}, e.g. in case of two air coils with an equivalent circuit model as shown in figure 5.2 with

$$L_{test} = L_1 + L_2 + 2M_{mag} \tag{5.9}$$

and the sum of the parasitic resistances R_{test} of these air coils with

$$R_{test} = R_1 + R_2 \tag{5.10}$$

to

$$\underline{Z}_{test}(f) = R_{test} + j2\pi f L_{test} \tag{5.11}$$

and

$$\underline{U}_{PoC}(f) = \underline{Z}_{test}(f)\underline{I}_{PoC}(f) + \underline{U}_g(f). \tag{5.12}$$

Depending on the wiring of the air coils, the different combinations of parallel and series connections between different air coils define L_{test}. There are always two air coils physically built together, e.g. called a block of air coils. If wired together (electrically connected), their magnetic coupling affects the overall test stand impedance as well. [10]
Additional magnetic coupling results from the air gap between different blocks of air coils that can be varied in distance. The option to either change the air gap between different blocks of air coils and to further connect multiple blocks together allows the stepwise large increase of the test stand inductance by 1 mH - 1.3 mH (individual inductance of one air coil), 2.5 mH - 3.5 mH (both inductances wired in series) or 0.45 mH - 0.7 mH (both

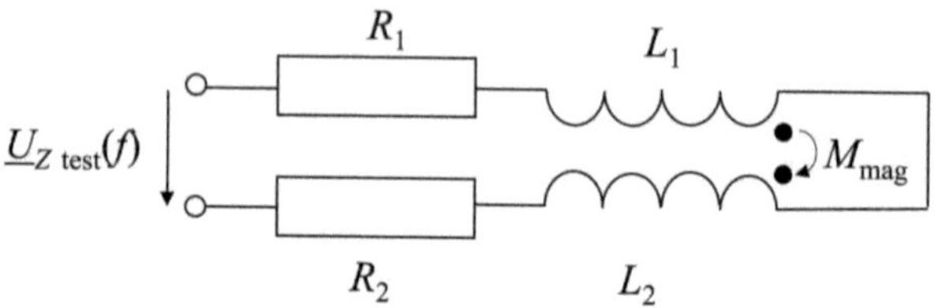

Figure 5.2: Electric circuit model of two air coils based on [10].

inductances wired in parallel). Each air gap allows a smooth variation of up to 1 mH of the overall test inductance. For the setup of the three test cases, three blocks of air coils have been used and the aggregated resistance of connection wires and air coils has been identified with R_{test} of 0.7 Ω.
The chosen inductances values to be set by the air coils are listed in table 5.1. Compared to typical inductance values of public LV networks, the values for test case 2 and test case 3 are much larger though the intention of the measurements is to measure the instability and not the characteristics of the inverter operation in typical LV networks. Typical inductance values in public LV networks have been identified by the author below 600 µH though especially when fitting the resonance characteristics to an RL-equivalent, the equivalent inductance value of the network can increase significantly.

Table 5.1: Test case inductances [10].

Test case	1	2	3
Applied inductance	0 mH	2.3 mH	3.2 mH

The test cases are depicted in figure 5.3 and the variation of the network impedance (solid lines in blue, cyan and red) becomes visible. In this section, the measurements have been performed for an operating power of the inverter of 1.4 kW (see inverter impedance in figure 5.3). The test stand has been set up as depicted in figure 5.1 and the currents at the PoC are recorded for a sinusoidal background voltage $\underline{U}_{\text{g}}(f)$ with an amplitude of $\sqrt{2} \cdot 230$ V at 50 Hz as presented in the following section.

5.2.2 Result analysis

The results of the scenario-specific measurements are depicted in figure 5.4.
For the analysis, it is of interest to identify the critical frequencies with regard to the currents and voltages at the PoC in the measurements. Compared to frequency domain, wavelet domain (time-frequency domain) is more suitable since the convolution of frequencies is presented in amplitude changes over time related to the respective frequency.

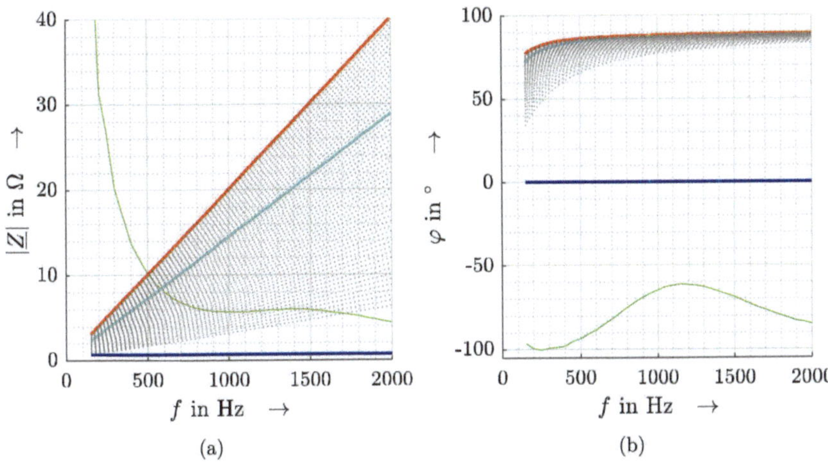

Figure 5.3: Magnitude (a) and phase angle (b) characteristics of impedance $\underline{Z}_{\mathrm{Inv}}$ of inverter I (green) and of test impedance $\underline{Z}_{\mathrm{test}}$ for test case 1 (blue), test case 2 (cyan) and test case 3 (red) based on [10].

An increasing distortion in the grid-side current results from increasing the network impedance. During the measurements of test case 3, the current response of the inverter causes its shut down after a certain time due to the too high voltage distortion resulting from the interaction of the inverter with the test stand impedance, see section 5.6 based on [22]. The respective wavelet domain plot is depicted in figure 5.4f. It can be visually identified that the critical frequency region for 3.2 mH is around 500 Hz.
For quantitative results, the current signal is studied in more detail, e.g. as presented in [15]. Still, the more detailed analysis identifies the same critical frequency region, e.g. around 500 Hz, for test case 3. In addition, the applied ESPRIT method [71] has identified a frequency shift of the dominant distortion component in the current spectrum of i_{PoC} from test case 2 with about 550 Hz to test case 3 with 500 Hz before the device shuts down. This is in accordance with the assumption that a higher network impedance leads to an impedance intersection at lower frequencies.
The theoretic conditions for which a stable operation is possible are defined clearly. In practice, parasitic and unconsidered effects as well as uncertainties and inaccuracies cause deviations from the theoretically stable operating conditions so that a minimum phase margin is required to enable a stable operation. Undocumented pre-test measurements have indicated these minor deviations between the theoretic analysis and laboratory measurements. A phase margin of at least 2 ° (instead of theoretically 0 °) has been required for a stable operation of the inverter. The origin of these minor deviations have not been identified but minor deviations, e.g. measurement uncertainties during the identification

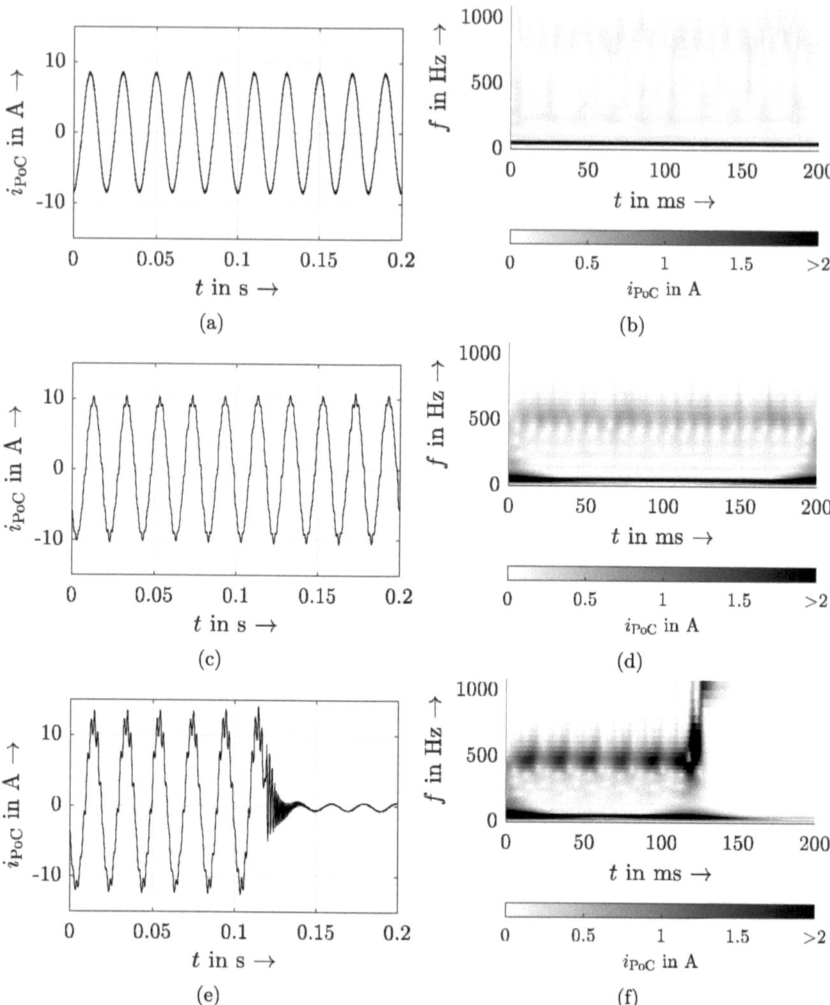

Figure 5.4: Grid-side current i_{PoC} in time domain (a, c, e) and wavelet domain (b, d, f) for test case 1 (a, b) test case 2 (c, d) and test case 3 (e, f) at operating power of 1.4 kW [10].

of the inductances of the air coils and for the voltage and current measurements, e.g. see appendix B, are not included in the theoretic analysis but can have an impact in practice.

5.3 Operating-point dependency

The previous measurements, e.g. as presented in figure 5.4, have been performed with a different measurement setup compared to the measurements in this section and the later sections. The inverter, i.e. inverter I, has been wall-mounted for the previous meausurements in section 5.2. For this section and the studies in the later sections, inverter I has been put on a moveable trolley and connected to the bus bar by additional cables. The inverter impedance has been measured at the PoC thus the connectors and cables have been included in the inverter impedances at the individual operating points resulting in different critical network impedances.
Based on the identified change of the inverter impedances at different operating powers, the previous measurements have indicated smaller impedance magnitude values for higher operating powers in the frequency region where the control has a relevant impact on the overall impedance of the inverter.
To analyze the impact of different operating powers and to validate the assumption that lower inverter impedance magnitudes are more critical, the impedance characteristics of inverter I in combination with different test impedances have been studied and published by the author in [19] as presented in the following of this section.

5.3.1 Test cases

Three different operating powers are studied in [19], i.e. 1 kW (small operating power), 2.5 kW (operating power in the middle of the possible range) and 4.5 kW (approximately maximum operating power - rated power). The list of applied inductances for the respective test cases is presented in table 5.2. Similar to the previous section 5.2, the intention of the measurements is to measure when the inverter becomes instable and not the characteristics of the operation in typical LV networks.

Table 5.2: Overview of applied inductances during test cases based on [19].

Test case n°	L_{test} at 1 kHz	Test case n°	L_{test} at 1 kHz	Test case n°	L_{test} at 1 kHz
1	No intended L	5	2380 µH	9	3785 µH
2	1075 µH	6	2385 µH	10	3943 µH
3	1625 µH	7	2600 µH	11	4112 µH
4	1915 µH	8	3500 µH		

Test case 6 has been added after the initial test case development to analyze test case 5 at 4.5 kW in more detail.
The test cases can be studied analytically. The theoretic results for the critical frequency, i.e. the impedance intersection, together with the respective phase angles are listed in table 5.3.

Table 5.3: Theoretic test case analysis [19].

Test case n°	Impedance intersection (f_{crit})	L_{test} at 1 kHz	$\varphi_{Z\,Inv}$	φ_{ZT}	φ_{PM}
	4.5 kW				
4	619 Hz	1915 µH	-93.1	82.3	4.6°
5	568 Hz	2380 µH	-95.2	83.2	1.6°
6	568 Hz	2385 µH	-95.2	83.2	1.6°
	2.5 kW				
9	470 Hz	3785 µH	-94.8	84.9	0.3°
10	462 Hz	3943 µH	-95.2	84.8	0°
11	453 Hz	4112 µH	-95.5	85.1	-0.6°
	1 kW				
10	465 Hz	3943 µH	-93.4	84.8	1.8°
11	455 Hz	4112 µH	-93.8	85.1	1.1°

5.3.2 Measurement results

An overview of the measurement results is listed in table 5.4.
It can be concluded that not only the network impedance but also the operating point of the inverter, i.e. the operating power, affects the inverter stability. The results validate the harmonic black-box stability assessment as both, the theoretic and the measurement-based analysis have indicated the same outcomes of the test cases.
In detail, the inverter tried to start multiple times before reaching steady state conditions for test case 5 at an operating power of 4.5 kW. As mentioned earlier, test case 6 has been added to determine the inverter operation and the interation with the test stand impedance in more detail. In test case 6, the inverter did not operate stable and shut down, indicating an insufficient phase margin. For higher test cases, i.e. test case 7 to test case 11, the inverter was not even able to start and synchronize with the voltage at the PoC.

Table 5.4: Measurement results for different operating powers [19].

Test case n°	Operating power		
	1 kW	2.5 kW	4.5 kW
1	stable	stable	stable
2	stable	stable	stable
3	stable	stable	stable
4	stable	stable	stable
5	stable	stable	stable, multiple attempts to start
6	not tested	not tested	instable
7	stable	stable	not started
8	stable	stable	not started
9	stable	stable	not started
10	stable	instable	not started
11	instable	not started	not started

The results for measurements at 1 kW, 2.5 kW and at 4.5 kW are depicted in figure 5.5 exemplarily. In figure 5.6, the results at an operating power of 4.5 kW are presented in time domain for test case 1 and test case 6. In test case 10, the inverter shuts down at an operating power of 2.5 kW though the inverter was able to reach a steady-state operating point for a short time. In test case 11, the inverter was not able to start as previously also detected for 4.5 kW at test case 6. For an operation at 1 kW, the inverter was still able to operate for test case 10 but operated instable for test case 11. All measurements have been measured at least twice, in many test cases more than three times, to ensure that the measurement results are reproducible. [19]

5.3.3 Theoretic application

The previous section 5.3.2 has validated the operating point dependency. For a more general assessment of the operating point dependency, the impedances of the simulation models (inverter A - D) and the measured inverters (inverter I - III) are studied theoretically with regard to their harmonic stability and their operating power.

Based on the measurements and the relation between highly inductive networks and the harmonic stability of the inverter, the critical inductance L_C and the critical frequency f_C are introduced by the author. The critical inductance represents the lowest inductance value for which a purely inductive network causes the inverter to shut down. For this inductance value, the amplitude and the phase margin criterion are violated while the critical frequency indicates the respective critical frequency region.

The resistive network component is neglected (i.e. no damping) so that the assessment

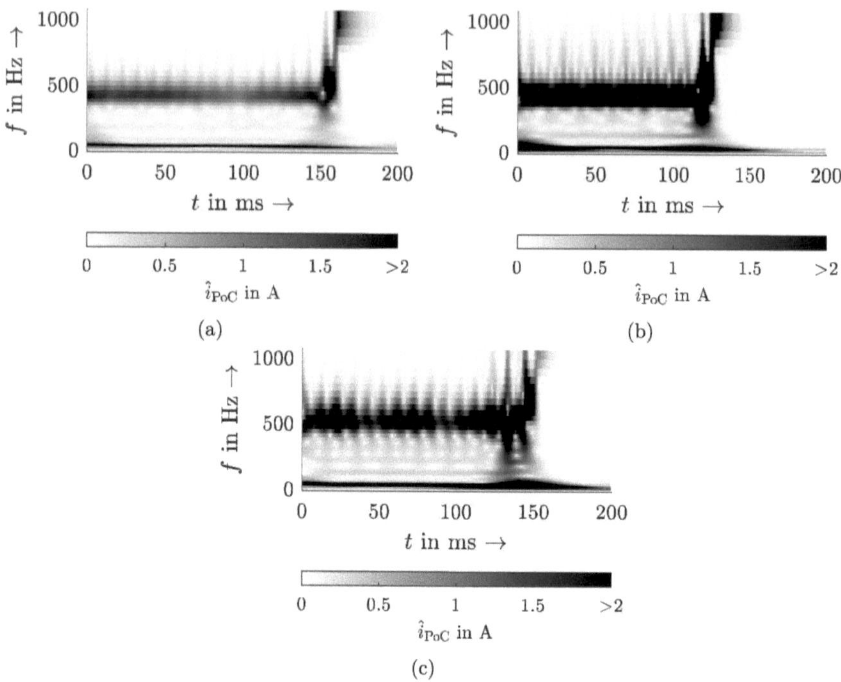

Figure 5.5: Grid-side current i_{PoC} in wavelet domain für 1 kW at test case 11 (a), 2.5 kW in test case 10 (b) and for 4.5 kW for test case 6 (c) [19].

follows a conservative approach. In practice, this can occur for networks with a series voltage regulator (SVR) and a high penetration of PE devices.

The analysis of the critical inductance and the critical frequency is performed in two steps [19] for each operating power of each inverter for any possible inductance:

1. It is analyzed, if the intersection of the magnitudes of the network impedance and the inverter impedance is in the active frequency region of the inverter impedance.

2. The phase margin at the frequency of the intersecting impedance magnitude is calculated.

The network inductance value is increased for each time the two steps are performed, until the first critical inductance value has been found that violates both, the amplitude and the phase margin criterion.

The calculated, critical inductance values and critical frequencies for all seven inverters are presented in figure 5.7.

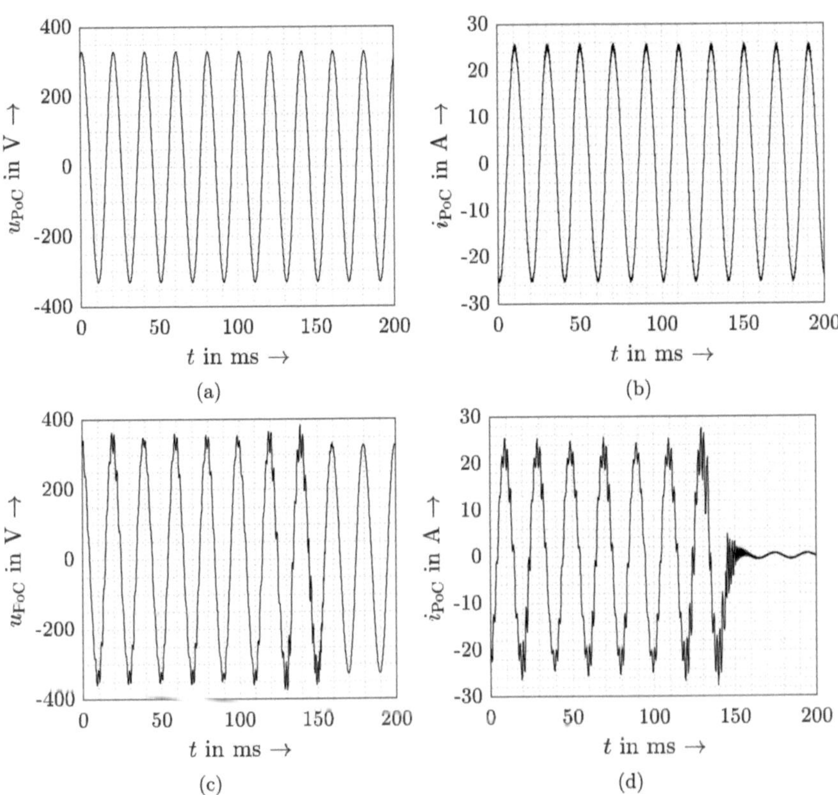

Figure 5.6: Voltage $u_{PoC}(t)$ (a,c) and current $i_{PoC}(t)$ (b,d) at the PoC in time domain für 4.5 kW at test case 1 (a, b) and test case 6 (c, d) [19].

With regard to the operating power, a shift in the critical frequency can be noted. This shift is inverter-specific and based on the inverter impedance characteristics (resonances, active frequency region and operating point dependency). Inverter B (figure 5.7, pink „□“) changes from 1.1 kHz at 0.5 kW to 1450 Hz at 4.5 kW, i.e. a shift by 300 Hz. As another example, inverter I (figure 5.7, green „∇“) changes from 500 Hz at 1 kW to 650 Hz at 4.5 kW, e.g. by 150 Hz. Not only the shift is inverter-specific, but also the absolute values of the critical frequencies. For inverter II (figure 5.7, cyan „▷“), the lowest critical frequencies are already below 400 Hz. On the other hand, for inverter B (figure 5.7, pink „□“), the lowest critical frequency is above 1.1 kHz. Many of the studied inverters have a non-monotonous change of the critical inductance and the respective critical frequency related to an increasing operating power [19]. For inverter III (figure 5.7, chestnut „◁“), the active frequency region is only present in the low frequency region for an operating

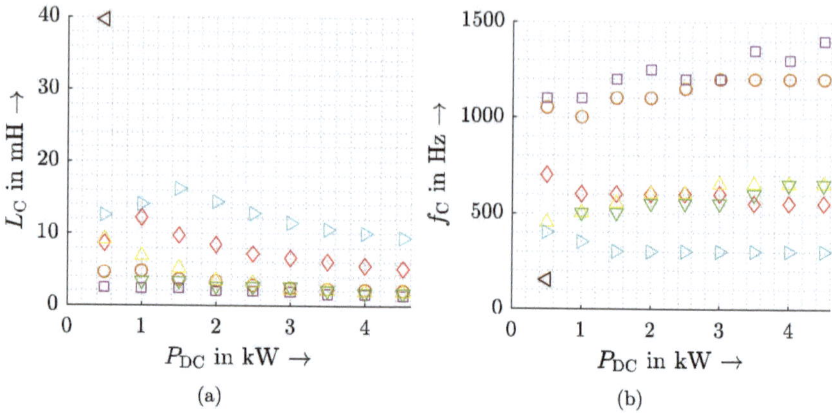

Figure 5.7: Critical inductance L_C (a) and critical frequency f_C (b) for inverter A (orange „○") inverter B (pink „□") inverter C (magenta „◇"), inverter D (yellow „△"), inverter I (green „▽"), inverter II (cyan „▷") and inverter III (chestnut „◁") based on [19].

power of 0.5 kW which leads to a very high critical inductance so that this inverter operates particularly stable.

5.3.4 Impact of phase angle of inverter impedance

The frequency-dependent phase angle characteristics of the inverter result from the components of the inverter and their specific parameters. The passive components, e.g. the AC-side filter circuit, and the active inverter components, e.g. the DC-link voltage control, the AC-side current control and the PLL, affect the impedance characteristics. Specifically the control is power-dependent. For high operating powers, the control dominates the impedance within its bandwidth. For low powers on the other hand, the impact of the control is smaller, e.g. due to smaller reference signals and smaller injected powers, while the impact of the AC-side filter is relatively higher on the inverter impedance characteristics. Therefore, the phase angle of the inverter impedance is more active for higher operating powers than for lower operating powers.

As a conclusion regarding the dependency on the operating point, i.e. the operating power, it becomes clear that not only one operating point but a representative number of operating points, e.g. in the full range of possible operating powers, need to be included for an overall statement towards an inverter-specific generalized harmonic stability.

5.4 Impact of resonances in the network impedance

To be more specific with regard to the previous sections, the analysis consideres rather frequency-dependent magnitudes and phase angles of the network impedance than equivalent circuit elements, e.g. inductance values. The electric (equivalent) network that represents the network impedance magnitudes and phase angles is not relevant. The network impedance magnitudes and phase angles do not necessarily require high inductance values but can also result from locally high network impedance magnitudes, e.g. due to resonances.

Real LV networks can have resonances in their impedance characteristics which consequently have to be considered in the analyses. In [72], multiple condensators have been connected to an LV network to study their impact on the network resonance frequency. The study concludes that the more condensators (or more precisely the larger the capacitance) in the network, the lower the resonance frequency and the larger the resonance amplification. However, [72] has only studied the magnitude characteristics of the network impedance though the phase angle characteristics are important as well. A similar effect was measured in [34] where the authors have reconstructed an instability based on the same phenomenon.

The impact of an exemplary network impedance resonance is published by the author in [6] and presented in the following of this section accordingly. To demonstrate the impact of network impedance resonances compared to neglecting the resonances when extrapolating the respective RL-equivalent identified at power frequency, i.e. 50 Hz, a suitable network impedance and inverter impedance design have to be chosen.

5.4.1 Simulation model configuration

In the following, the specific simulation model for the analysis of the impact of the resonance is explained briefly.

Inverter design

To study the relevance of the resonances in the LV network impedance, an inverter is designed with not optimal parameter settings. With the implemented AC-side filter parameters of the LCL-filter I of the modular model as disclosed in appendix A.1 and figure A.1, the resonance frequency of the LCL-filter $f_{LCL\text{res}}$ can be calculated in terms of

$$f_{LCL\,\text{res}} = \frac{1}{2\pi}\sqrt{\frac{L_{\text{f d}} + L_{\text{f g}}}{L_{\text{f d}} L_{\text{f g}} C_{\text{f}}}} \tag{5.13}$$

to 3.07 kHz.

The resulting inverter impedance characteristics are depicted in the later depicted figure

5.9 for the operating power of 3.9 kW that is set for the simulation.

Low voltage network design

For the design of the LV network impedance in terms of the interaction with the inverter, four theoretic cases are considered: a stable operation of the inverter (test case 1), a critical operation (marginally stable) of the inverter (test case 2), an operation for an LV network with a resonance in the impedance with a magnitude value at power frequency as in test case 1 (test case 3) and an instable operation of the inverter (test case 4). The critical impedance (test case 2) can be calculated based on the intersection of the inverter impedance magnitude and the network impedance magnitude as introduced earlier and in [19]. For the RL-equivalent of the network, the intersection of both impedance magnitudes shifts in the frequency range depending solely on the RL values, if the inverter impedance characteristics remain unchanged. Figure 5.9 (b) shows that the phase angle of the inverter impedance drops again slightly below 90° for the frequency of 1050 Hz. To cause an intersection, the inverter impedance magnitude is 21 Ω at this frequency. To fit the respective RL-equivalent of the network impedance with a negligible small resistance, the inductance value is calculated with

$$L_\mathrm{C} = \frac{21\,\Omega}{2\pi\,1050\,\mathrm{Hz}} \approx 3.2\,\mathrm{mH}. \tag{5.14}$$

The inductance values for the stable test case (test case 1) are chosen below and for the instable test case above the critical inductance value. Table 5.5 lists the network element parameters for the individual test cases.

Table 5.5: Test case network parameters based on [6].

Test case	Network type	Equivalent ciruit model	Parameter	Value	Parameter	Value
1	$\underline{Z}_{\mathrm{g\,stable}}$	figure 5.8 (a)	R_g	0.07 Ω	L_g	2.9 mH
2	$\underline{Z}_{\mathrm{g\,crit}}$	figure 5.8 (a)	R_g	0.07 Ω	L_g	3.2 mH
3	$\underline{Z}_{\mathrm{g\,res}}$	figure 5.8 (b)	$R_{\mathrm{g}2}$	0.02 Ω	$L_{\mathrm{g}2}$	2.6 mH
			$R_{\mathrm{g}1}$	0.05 Ω	$L_{\mathrm{g}1}$	0.3 mH
			R_C	0.055 Ω	C_g	45 µF
4	$\underline{Z}_{\mathrm{g\,instable}}$	figure 5.8 (a)	R_g	0.07 Ω	L_g	5 mH

The respective electric network models are depicted in figure 5.8.
The impedance characteristics for the different test cases and the inverter are presented in figure 5.9. The network impedances of the RL-equivalent circuit models are highly inductive. The network impedance with the resonance has a capacitive phase angle characteristic around the resonance frequency.

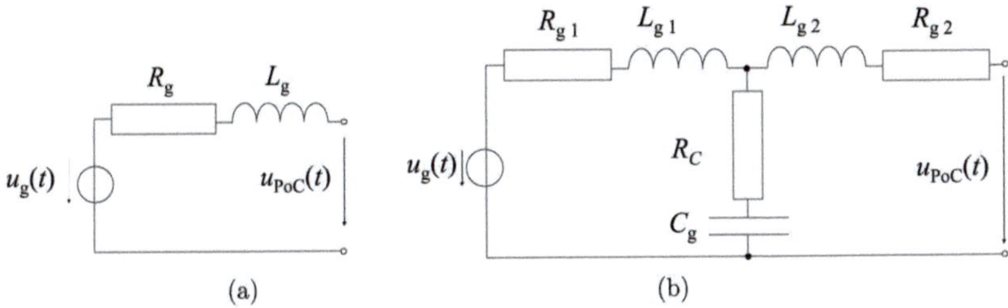

Figure 5.8: Equivalent *RL*-circuit model of an LV network (a) and a resonance network LV network (b) based on [6].

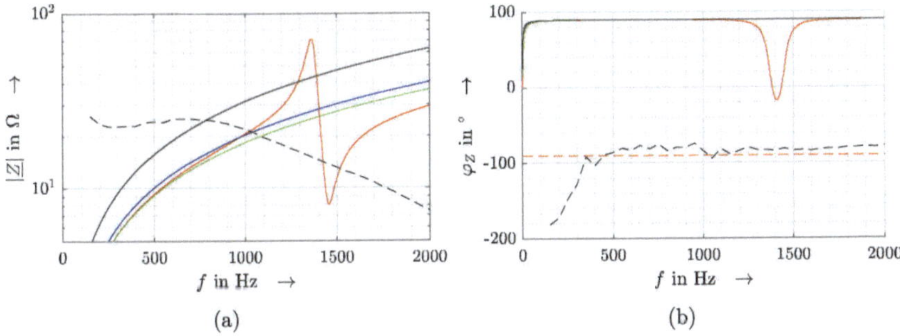

Figure 5.9: Magnitude (a) and phase angle (b) characteristics of the network impedance $\underline{Z}_g$ for test case 1 (green), test case 2 (blue), test case 3 (red) and test case 4 (black) and inverter impedance $\underline{Z}_{Inv}$ (black-dashed) and $-90°$ line (red-dashed line) in subfigure (b) based on [6].

5.4.2 Simulation results

To excite the entire system and to trigger an interaction between the inverter and the network impedance, a 250 Hz component (5th order harmonic) with an amplitude of 9.75 V is superimposed as a step change at the beginning of the simulation to the sinusoidal voltage at 50 Hz with an amplitude of 325 V. The 5th order harmonic is often present in the spectrum of the background voltage of public LV networks and therefore chosen for step change. The current response for all four test cases is recorded and depicted in figure 5.10.

The larger network inductance in test case 2 compared to test case 1 destabilizes the system. The current at the PoC is only marginally stable (see figure 5.10 (b)). The current response is strongly distorted. For test case 4, the instable operation of the inverter is presented in figure 5.10 (d). Though the simulation makes a theoretic response of the inverter possible, the overcurrent protection of the inverter would have been triggered in practice to prevent the damage of hardware components of the inverter.

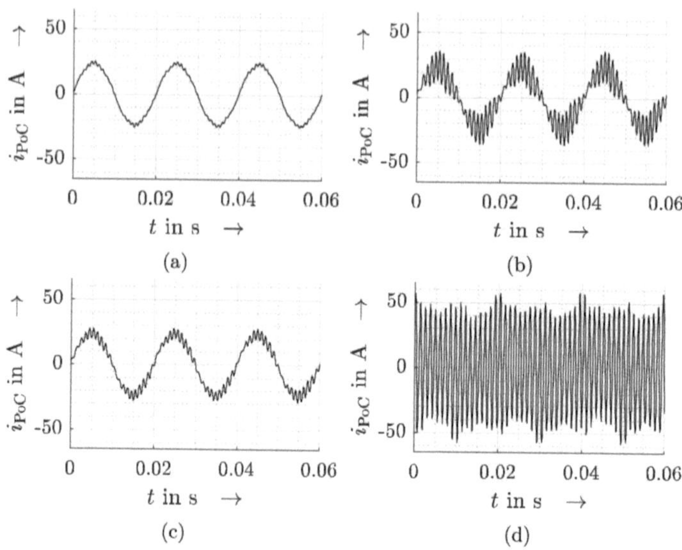

Figure 5.10: Current i_{PoC} at PoC for test case 1 (a), test case 2 (b) test case 3 (c) and test case 4 (d) based on [6].

For test case 3 as depcited in figure 5.10 (c), the distortion is much higher than in figure 5.10 (a). Though both test cases have the same R/X-ratio at power frequency, i.e. 50 Hz, the resonance in test case 3 causes the high current distortion around 1 kHz.

A discrete Fourier transform (DFT) has been performed on the current at the PoC to identify the dominant frequencies in the current distortion and the respective current amplitudes.

Though the step change excites the entire spectrum only once at the beginning of the simulation, the constant oscillation of the current at the PoC is visible for the test cases 2-4 in figure 5.10. This one-time excitation cannot be dampened in the critical frequency range where the inverter impedance magnitude and the network impedance magnitude is similar (see figure 5.9) and remains present during the simulation. These frequency ranges are for test case 1 around 1.1 kHz, for test case 2 and test case 3 around 1 kHz and for test case 4 around 800 Hz. On the other hand, the constant voltage excitation at 250 Hz does not cause a dominant current response. In figure 5.11, oscillations in the frequency range around 1.1 kHz become also visible for test case 1. However, these oscillations are negligible considering the logarithmic scale of figure 5.11 and consequently the irrelevant share in the current spectrum though indicating the resonance at this frequency but not an amplification.

It can be concluded that resonance amplifications challenge the operation of the inverter.

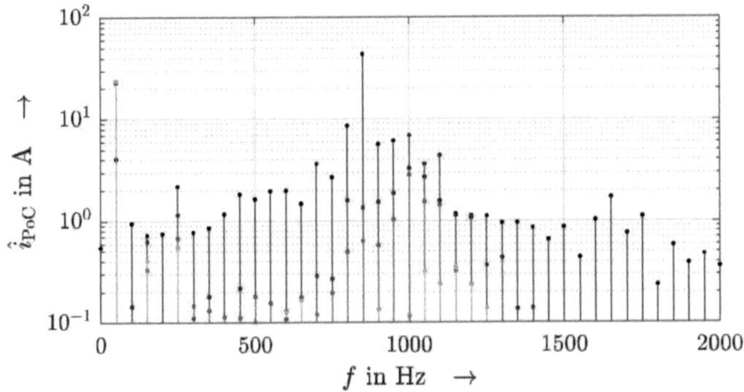

Figure 5.11: Frequency spectrum of current i_{PoC} at the PoC for test case 1 (green), test case 2 (blue), test case 3 (red) and test case 4 (black) scenario based on [6].

Due to a strong resonance amplification, an intersection of the network impedance magnitude and the inverter impedance magnitude is more likely [10]. The phase angle of the network impedance drops typically around the resonance frequency. As mentioned previously, highly inductive networks are more critical for the inverter but an unfavorable phase angle characteristic in combination with a high resonance amplification can cause critical interactions of the inverter with the network impedance as well. Only considering the (extrapolated) R/X-ratio at power frequency is not sufficient, if resonances in the network impedance are present.

5.5 Probabilistic consideration of network impedances

The previous section has demonstrated the impact of resonances in the LV network impedance on the inverter stability. In practice, these resonances can vary significantly. The large number of possible network configurations and the respective aggregated network impedances at the PoC of the inverter challenge assessment strategies of selected and individual LV network configurations. As a further development of the harmonic stability analysis, it is suitable to use real measured network impedances of public LV networks, instead of reference impedances or artificially created study cases. The impact of real measured network impedances on the harmonic stability of commercially available inverters has been studied and published in [1] by the author and is presented in the following of this section.

With regard to a more practical approach, only commercially available devices are analyzed. Their impedance characteristics have been identified in the laboratory based on the FS as introduced earlier (reference voltage with an amplitude of 325 V at 50 Hz and a sweep component with an amplitude of 5 V, a phase angle $\varphi_{\text{PoC}\,U}$ of 0 ° in relation to the

50 Hz component and 50 Hz steps as sweep parameters).
Inverter III has not been physically available for the measurements at the time when this study was performed and has previously only been identified up to 2 kHz. On the other hand, other inverters have been included so that the number of inverters is extended to six inverters, i. e. four more commercially available inverters. One of the four inverters, i.e. inverter IV, is a three-times single-phase inverter (three individual single-phase modules that are connected phase-neutral), the other three inverters, i.e. inverter V, inverter VI and inverter VII are true three-phase inverters. The three-phase inverters are studied based on their single-phase equivalent circuit model to apply the same harmonic stability assessment method as previously introduced. The limitation is that the overall system configuration is assumed rather balanced. The inverter impedances can be assumed balanced due to their symmetric implementation. However, the assumption regarding balanced network impedances and their impact on the reliability of the stability statement is left to future work. Consequently, the results regarding the three-phase inverters are only to gain a first perspective on future work in terms of the inverter impedances in relation to LV network impedances as the applicability on three-phase inverters has to be proven in practice by measurements.
Based on the previous finding that the inverter impedance is power-dependent, the following analysis considers the inverter impedances at 10 % and at 100 % of the rated power as depicted in figure 5.12 and [1]. The network impedances and the respective measurement method are disclosed in appendix C.2.

5.5.1 Probabilistic method

A scheme of the probabilistic method is presented in figure 5.13. To compare the magnitudes at similar frequencies according to the amplitude criterion (intersection of the impedance magnitudes), the discrete impedance magnitudes are interpolated to match the inverter impedances at the measured frequencies.
For the analysis, three loops can be performed to evaluate the data. The outer loop iterates trough all inverters (first loop, counter i). Each inverter has to be compared to all measurement sites (second loop, counter g) and the inner loop finally compares each inverter at each measurement site for a specific frequency (third loop, counter f).
During each iteration step, it is tested if the impedance magnitudes of an inverter and the respectively studied measurement site intersect. If an intersection occurs, the phase margin criterion is applied, or, if no intersection occurs, the frequency counter or respectively the measurement site counter and possibly the inverter counter are increased.
However, if the phase margin criterion is applied, i.e. due to the intersecting impedance magnitudes, and indicated a violation of the Nyquist criterion, the combination of the measurement site and the inverter are saved together with the respective critical frequency.

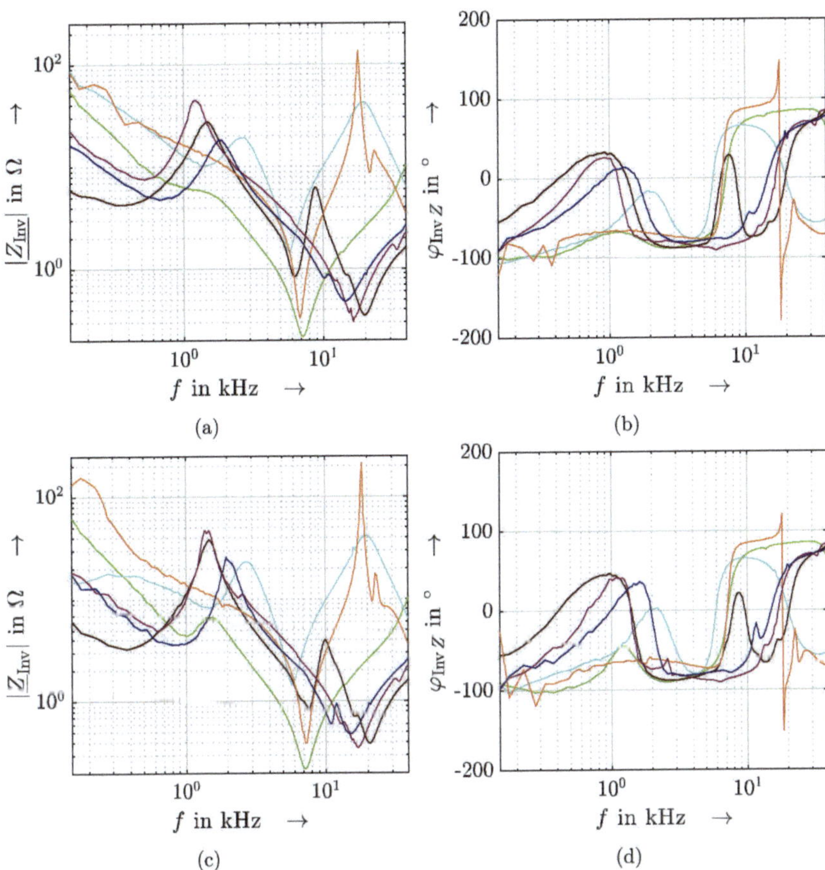

Figure 5.12: Characteristics of impedance magnitudes $|\underline{Z}_{\mathrm{Inv}}|$ (a) and phase angle $\varphi_{\mathrm{Inv}\,Z}$ (b) of six commercially available inverters (inverter I (green) , inverter II (cyan), inverter IV (brown), inverter V (orange), inverter VI (purple) and inverter VII (blue)) at 10 % (a, b) and of 100 % (c, d) of rated power based on [1].

5.5.2 Measurement-site specific results

In detail, figure 5.14 presents the intersections of the impedance magnitudes sorted for the inverters.

It has been found that no combination of any inverter with any measurement site indicates a violation of the phase margin criterion despite the large number of intersecting impedance magnitudes.

However, in practice, the network impedance does not remain the same but is time de-

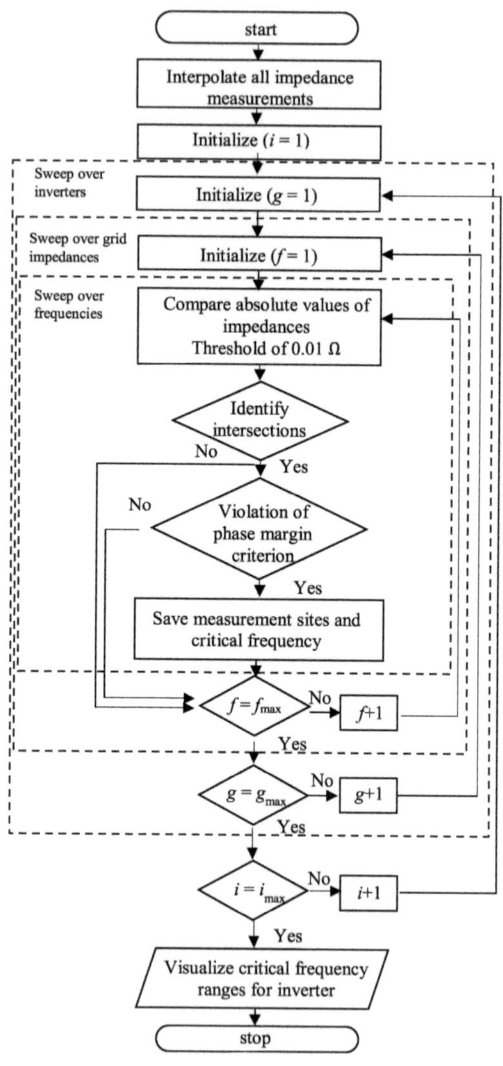

Figure 5.13: Scheme of the method to evaluate the probabilistic stability [1].

pendent (hourly, daily and seasonal). Therefore, the author proposes to adapt the phase margin criterion and add a reserve of 30 ° to ensure the stability of an inverter by including an uncertainty of the network impedances. This value is based on discussions of the

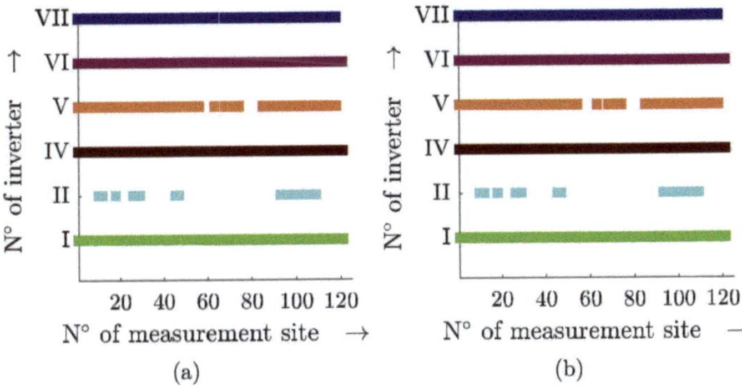

Figure 5.14: Combination of measurement sites and inverters at 10 % (a) and 100 % (b) of the rated power with intersecting impedance magnitudes based on [1].

author with manufacturers that apply a similar reserve. Possibly, the quantitative value, i.e. 30 °, has to be refined in the future based on more detailed studies of the fluctuation of network impedances in public LV networks. For the following,

$$\varphi_{\mathrm{PM}} > 30\,° \tag{5.15}$$

is considered for the phase margin criterion as an indication of a stable system.

The combinations of measurement sites and inverters that violate the stability criterion with the additional phase margin of 30° are presented in figure 5.15.

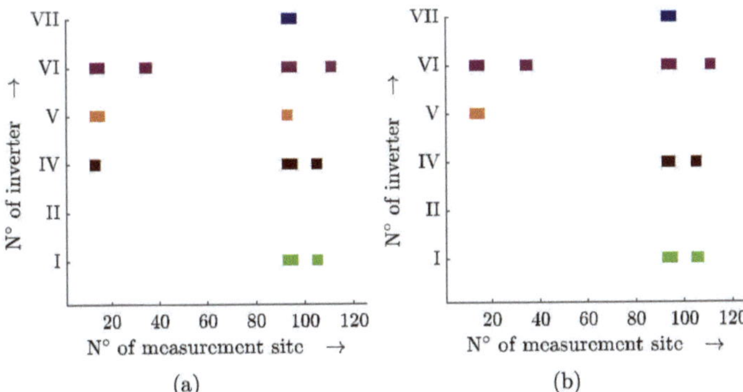

Figure 5.15: Combination of measurement sites and inverters at 10 % (a) and 100 % (b) of the rated power with intersecting impedance magnitudes and critical phase margin based on [1].

The results indicate measurement sites at which the operation of an individual inverter is

challenged for all inverters except for inverter II. Furthermore, when both the impedance magnitude intersection and the phase margin criterion are studied, it becomes obvious that only certain measurement sites are critical while a large number of measurement sites does not challenge the inverter operation of any of the inverters. Besides evaluating the combinations of measurement sites and inverters, the specific frequencies that are of relevance, either for the individual inverter, or a specific measurement site, can be evaluated.
The critical frequencies for the inverters are depicted in figure 5.16.

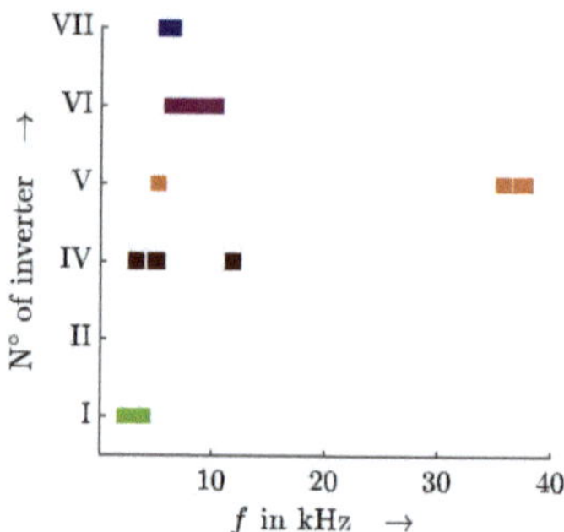

Figure 5.16: Inverter-specific critical frequencies, studied at 10 % and 100 % of the rated power of the inverters, related to the set of studied measurement sites based on [1].

Studying the measured frequency range (up to 39 kHz) with focus on individual inverters, it becomes obvious that only certain frequency regions are critical. Exemplarily, figure 5.17 depicts the critical frequencies of inverter VI and the respective phase angles at these frequencies. As explained earlier, the phase angle of the inverter impedance is below 90 ° at all critical frequencies.
These frequency regions can be narrowed down to one up to four regions of interest while a large range of frequencies does not cause problems. These frequency regions with critical frequencies are inverter specific. This shows again that the different inverter designs and inverter characteristics cause inverter-specific challenges. Knowing the critical frequency regions of the inverter for all operating powers, the manufacturer can implement measures to adapt the design more punctual and efficient as briefly addressed in the later section 5.8.
The critical frequencies can also be studied with regard to the measurement sites. Also for the measurement sites, the critical frequencies are narrowed down to a small number and a narrow ranges. Only few measurement sites challenge the inverter operation in the high-frequent range, i.e. between 35 kHz and 39 kHz (maximum evaluated frequency) around the second band of the effective switching frequency. For the distribution system operator (DSO), the frequencies that are challenging for inverters can be of interest,

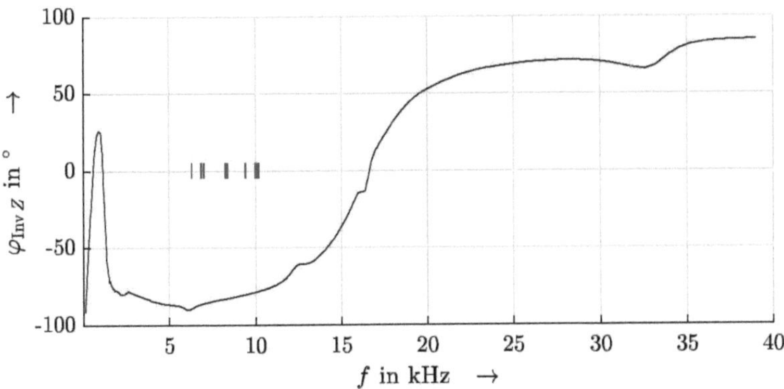

Figure 5.17: Exemplary phase angle characteristic φ_{Inv} (purple) of inverter impedance at 10 % of the rated power of inverter VI and critical frequencies (red) for all operating powers (10 % - 100 %) indicated on the abscissa [1].

especially with focus on higher penetration levels of PV systems but also electric vehicle (EV) chargers in households and battery storage systems (BSSs) that can be analyzed similarily. Consequently, the DSO could also identify the grid compatibility of an inverter with respect to a particular power grid and a specific PoC, i.e. the measurement site, where a device is installed.

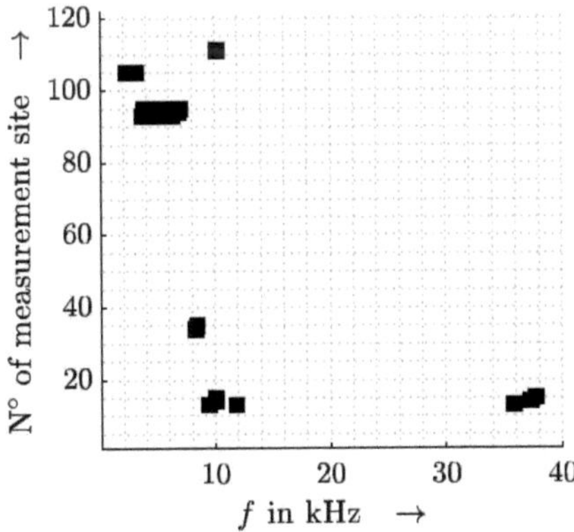

Figure 5.18: Critical frequencies at individual measurement site related to the set of studied inverters based on [1].

5.5.3 Grid compatibility index

For a quantification of the grid compatibility, the gci is introduced in [1]. The index relates the number of critical measurement sites n_{crit} to the number of studied measurement sites n_{tot} for one specific inverter in terms of

$$\text{gci} = 1 - \frac{n_{\text{c}}}{n_{\text{tot}}}. \tag{5.16}$$

The closer the index to one, the better is the grid compatibility with regard to the studied set of measurement sites. A gci of one indicates no stability issue for the respective inverter at any of the studied measurement sites. The gcis of the studied inverters are listed in table 5.6. Inverter II is not expected to show harmonic stability issues.

Table 5.6: Grid-compatibility index gci [1].

Inverter	I	II	IV	V	VI	VII
Grid compatibility index gci	0.9669	1.000	0.9587	0.9669	0.9256	0.9835

Next to assessing the grid compatibility, the gci can be used as an optimization criterion. If the inverter design is varied, the change of the gci can indicate, if the adaption of the inverter design has improved or worsened the grid compatibility. In future, it is possible, to consider a frequency-specific gci for a more detailed assessment.

5.6 Limitations of the harmonic stability analysis

The previous sections of this chapter have applied control theory by classical definitions of stability. The general understanding is that a stable system provides damping to a signal excitation such that it decreases the exciting signal. In contrast, an instable system amplifies an exciting signal. With regard to the black-box stability analysis, the methods relate to the small-signal stability and are not applicable for large-signal studies. In the following of this section, the limits of this theory with regard to a stable inverter operation are addressed and have been published by the author in [22].
Besides the network impedance that has been studied in the previous analysis, the waveform of the background voltage $u_{\text{g}}(t)$ is non-ideal, i.e. non-sinusoidal, so that voltage frequency components next to the power frequency f_1 are present. Though the occurrences of voltage frequency components at the PoC are specified in standards, e.g. the EN 50160 [38], the detailed spectrum can vary largely in practice. Typical frequency spectra of the background voltage for a public LV network (so-called flat-top voltage) and an industrial LV network (so-called pointed-top voltage) have been disclosed in [53] and are listed in section C.1.

In practice, the shutdown is realized by either overcurrent and overvoltage protectors or via the grid-side IGBTs to disconnect the inverter from the public LV network. If the reason for the disconnection of the inverter cannot be related to formal stability analyses, this shutdown is called adverse behavior or tripping which includes also malfunctions of inverter components. Consequently, by definition of the author, stability is only one aspect of the stable operation of the inverter thus the formal stability analysis is not sufficient to assume a stable inverter operation.
With regard to classical stability theory, the background voltage distortion presents a disturbance to the system. If the system provides enough damping, e.g. as theoretically calculated, the system is considered stable. The background voltage distortion is consequently not included in the harmonic stability analysis since it does not affect the stability of the system by definition. However, strong signal excitations (steady state and transient) can also trigger the overcurrent and overvoltage protection, even in absence of a network impedance. Therefore, the impact of a background voltage distortion on the stable operation of the inverter is of importance but not part of harmonic instabilities. [22]

5.6.1 Measurement setup

To compare the inverter operation with a sinusoidal background voltage, a pointed-top voltage and a flat-top voltage, the same measurement setup with the air coils as depicted in figure 5.1 is used. The pointed-top and the flat-top voltage are applied with the voltage spectrum from [53] as listed in section C.1 and the inverter response, i.e. the current at the PoC is measured.
The same test cases as presented in table 5.2 are measured with regard to the network inductance, but three voltage waveforms are applied for each test case. The sinusoidal voltage is applied first. Directly afterwards, it is switched to the flat-top voltage and finally, the pointed-top is applied, all within one measurement.

5.6.2 Measurement results

The measurement results are listed in table 5.7.
For test cases where the flat-top voltage waveform has caused the inverter to shut down, the pointed-top voltage waveform has not been applied, since for theses test cases, the background voltage had already indicated the impact of its distortion. The pointed-top voltage waveform has not caused additional shut downs for those test cases where the flat-top voltage waveform allowed a stable operation of the inverter.

Table 5.7: Measurement results for sinusoidal and flat-top reference voltage [22].

Operating power	**Inductance value** at 1 kHz	**Operation status** Sinusoidal voltage waveform	Flat-top (and pointed-top) voltage waveform
1 kW	3785 µH	Stable	Stable
	3943 µH	Stable	Instable
	4112 µH	Instable	-
2.5 kW	3785 µH	Stable	Stable
	3943 µH	Instable	-
	4112 µH	Instable	-
4.5 kW	1625 µH	Stable	Stable
	1908 µH	Stable	Instable
	2385 µH	Instable	-

The time-domain data for the voltage and the current at the PoC is shown for a sinusoidal voltage waveform and an operating power of the inverter of 1 kW in figure 5.19 and for the flat-top and the pointed-top voltage waveform and the respective current in figure 5.20. For both measurements, a test stand inductance of L_{test} of 3.943 mH is applied. This represents the test case for an operating power of 1 kW where the inverter is able to operate stable with a sinusoidal background voltage but shuts down for the flat-top voltage waveform as visualized in figure 5.21 and figure 5.22.

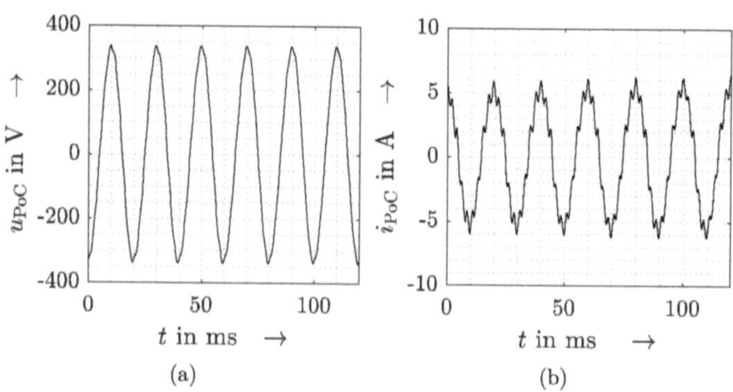

Figure 5.19: Voltage u_{PoC} (a) and current i_{PoC} of inverter I at an operating power of 1 kW with sinusoidal background voltage and a test stand impedance L_{test} of 3.943 mH [22].

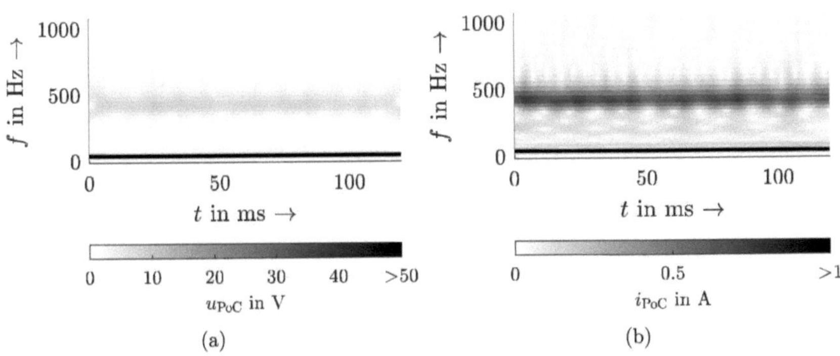

Figure 5.20: Voltage u_{PoC} (a) and current i_{PoC} of inverter I at 1 kW in wavelet domain with sinusoidal background voltage and a test stand impedance L_{test} of 3.943 mH [22].

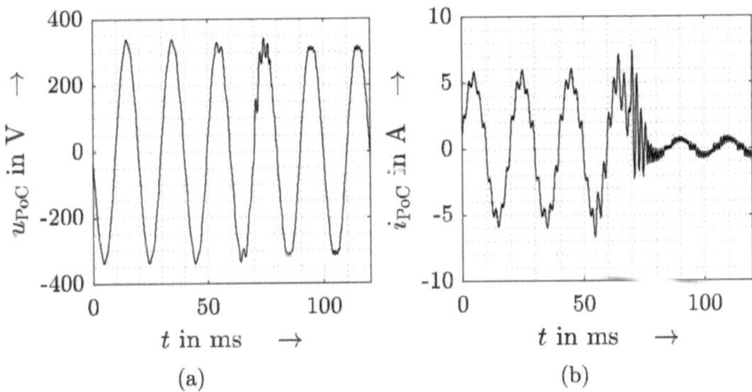

Figure 5.21: Voltage u_{PoC} (a) and current i_{PoC} of inverter I at 1 kW in time domain with flat-top and pointed-top background voltage and a test stand impedance L_{test} of 3.943 mH [22].

The results listed in table 5.8 can be categorized into stable (category 1) and instable (category 2) inverter operations. For the instable operation of the inverter, it can be further distinguished between the device shut down during steady state (category 2A), the device shut down during the change of the voltage waveform (category 2B) and the device was not even able to reach a steady state operation (category 2C) as listed in table 5.8 [22]. A stable operation of the inverter (category 1) has been measured for lower values of L_{test} than annotated for the respective operating powers in table 5.7 as already presented in section 5.3 and table 5.4 for sinusoidal voltage. These measurements have been repeated for the flat-top and the pointed-top voltage waveform. As they have not caused an instable inverter operation, they have not been listed in table 5.7.

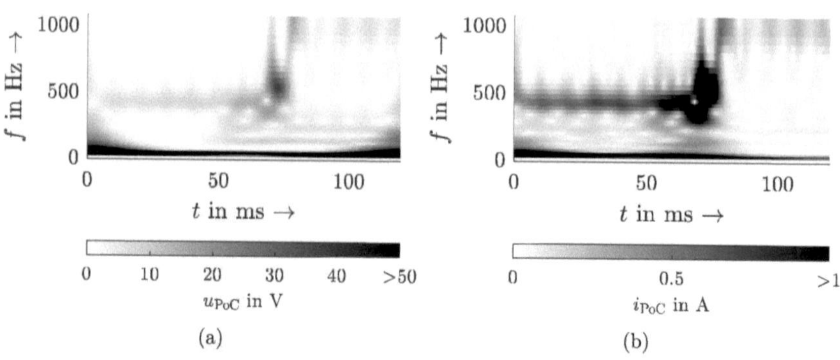

Figure 5.22: Voltage u_{PoC} (a) and current i_{PoC} of inverter I at 1 kW in wavelet domain with flat-top and pointed-top background voltage and a test stand impedance L_{test} of 3.943 mH [22].

Table 5.8: Categories of measured results [22].

Category		Stability statement	Type of instability	Background voltage waveform	Specifications
1		Stable	-	Sinusoidal Flat-top Pointed-top	-
2	A	Instable	Steady state	Sinusoidal Flat-top	-
	B	Instable	Transient		Shut down when changing the background voltage
	C	Instable	Unspecified	Sinusoidal	Not reaching the intended operating point before shutting down

Also, the instable operation of the inverter for a sinusoidal background voltage has already been introduced in section 5.3 for steady state (category 2A).

The background voltage distortion, tested with the flat-top voltage waveform, has an impact on the stable operation of the inverter (category 2A). The measurements show that a distorted but reasonable background voltage can cause the inverter to shut down even when the impedance-based analysis indicates no instability of the inverter. The differences in the theoretic analysis and the measurements do not result from inaccuracies, e.g. measurement uncertainties. The previous operation with sinusoidal background voltage has been predicted accurately enough by the theoretic analysis to state reliably towards the inverter stability.

The transient stability of the inverter can be challenged when the background voltage distortion is changing (category 2B). This can occur specifically in case the impedance-based stability criterion already indicates a low stability margin. The voltage transients can be seen as a broadband excitation that can cause a large current response in systems with low damping. Furthermore, the point on wave (PoW), e.g. when the transient occurs in relation to the voltage component at power frequency, can have a large impact on the maximum current value which is typically at 90°. The impedance-based stability analysis can indicate the stability of the inverter but the frequency-specific immunity and sensitivity of the inverter can challenge a reliable prediction towards an overall stable operation of the inverter.

Strongly distorted voltages at the PoC can prevent the inverter to start its operation and to reach a steady state (category 2C). From a white-box perspective, the PLL is possibly not able to synchronize with the voltage at the PoC if the distortion is too high thus also the MPPT might not work properly if the power flow from the DC-side to the AC-side is interrupted frequently. This category 2C is neither related to the steady state stability nor the transient stability.

To conclude the limits of the stability assessment by following Nyquist or also, though not part of this work, the Eigenvalue analysis, by means of the harmonic small-signal black-box stability analysis, the background voltage distortion is neglected. It has been shown, that a background voltage distortion will affect the overall stable operation of the inverter. Though the background voltage distortion will not cause an instability by definition of the term stability, the background voltage distortion can worsen the voltage and current distortion at the PoC. The test cases with an applied background voltage distortion have always caused the inverter to shut down for lower values of the test stand inductance L_{test} than with sinusoidal background voltage.

5.7 Framework for stable operation

The previous sections have indicated that the black-box small-signal analysis can be performed to identify the impact of the frequency-dependent network impedance, e.g. including resonances, on the inverter operation. Section 5.6 has shown that solely considering the network impedance is not sufficient to make a statement towards the stable operation of the inverter. It can only indicate sufficiently, when it does not operate stable. Consequently, this section addresses a more holistic framework to advance from solely studying the classical stability to analyzing the overall stable operation of an inverter based on [18].

5.7.1 Emission, Immunity and Stability

From a broader perspective, the interactions in power grids are determined by emission, immunity and stability. The first two are furthermore also standardized and regulated.

Emission

The emission specifies the individual injection, e.g. of a device or installation, into the power grid. In practice, the contribution assessment, i.e. the share of emission for non-ideal conditions in the field and the root cause are highly political topics and not part of this work. For LV networks, the emission of grid-connected devices has to comply with standards such as the IEC 61000-3-2 [59] and the IEC 61000-3-12 [73]. The respective tests in the laboratory are performed without any network impedance or rather test stand impedance, e.g. ideal conditions. For grid-connected devices, the disturbances that result from the emission of an individual device can exceed the immunity limits of the device itself and of other devices, even when complying with the standards. This is typical in case of pronounced resonances in the network impedance but not necessarily a stability phenomenon. [18]

Immunity

The immunity test levels, e.g. according to the IEC 61000-4-13 [74], specify the range of distortion for which the device is expected to operate (interference-free). The standardized test proceedure applies tests for the immunity test levels but does not assess the device-specific limits which are of interest when comparing the device performance and the grid robustness of individual device designs [18].

Stability

While the previous work has studied mainly the stability, the stability can be related to the immunity. If the current emission is increasing for an interaction with any network impedance the voltage distortion at the PoC increases as well. Consequently, the immunity limits of the device are exceeded eventually. For an overall assessment of a device in practice, it is consequently rather the overall device immunity than the stability that is of importance for the stable operation.

5.7.2 Impact factors

This section addresses a more holistic understanding of the stable operation of the inverter that includes also the classical stability as analysed in the previous sections.
In practice, the minimum operating power of commercially available inverters for PV

applications has been measured by the author between 5 % and 10 % of their rated power. However, this describes rather the availability of PV systems with regard to the solar irradiance but not the reliability [18] and is not further considered in this work.
The analysis of the stable operation is related to the immunity of the inverter for which both, the background voltage and the network impedance define the voltage at the PoC. Analyzing these two impact factors, the signal characteristics of the voltage and the current at the PoC can be categorized into steady state and transients. An overview of the impact factors is given in table 5.9.

Table 5.9: Overview of impact factors [18].

Category	Impact factor	Signal characteristics
A.1	Network impedance	Steady state
A.2		Transient
B.1	Background voltage	Steady state
B.2		Transient
C.1	Background voltage and network impedance	Steady state
C.2		Transient

A. Network impedance

1 Steady state

The impact of the network impedance has been analyzed in the previous sections of this chapter, i.e. section 5.1 - section 5.5, in detail for the steady state and relates directly to the classical stability.

2 Transient

This work has only assessed the steady state impact of the network impedance. Changes in the network impedance can also cause shutdowns of grid-connected devices in case of a pronounced current response. As an example, a strong transient reaction that has caused PV inverters to shut down in response to the inrush of transformers in a public LV network has been reported in Australia [35].

Changes in the network impedance can cause voltage distortions at the PoC that result from the current response of an inverter. In test proceedures, changes of the test stand impdance are currently not considered. For future laboratory measurement assessments, suitable step changes and frequency-dependent impedance characteristics, e.g. resonances (frequency and amplification) or more general the magnitude and phase angle characteristics of the test stand impedance have to be defined. [18]

B. Background voltage

The background voltage distortion affects the voltage at the PoC but is not part of a feed-back loop. Especially for stiff grids, but also for emissions resulting from other grid-connected devices that are independent from the voltage at their PoC, e.g. switching frequency emissions, the background voltage is not affected by the current injection of the inverter. Consequently, classical stability analysis does not consider the background voltage, e.g. in the impedance-based stability criterion. In thoses analyses, the background voltage is seen as an independent disturbance. However, if the system is stable itself, it will provide sufficient damping, such that a dynamic excitation decreases and constant excitations will lead to a steady state instead of an increasing current response.

1 Steady state

A constant background voltage distortion can cause current and voltage distortions at the PoC. For too strong voltage distortions at the PoC, the inverter can fail to synchronize with the power frequency. The intended operating point cannot be reached which challenges the MPPT. Individual inverters have shown different sensitivity characteristics, e.g. inverter II has shown a high sensitivity on even low-order harmonics (second and fourth order mainly) compared to neighbouring odd order harmonics.

2 Transient

Measurements of transient voltages have demonstrated their impact on the stable operation of the inverter and its current response [3, 4]. Exemplarily, the current response of two different inverters, inverter I and inverter II, have been measured related to large step changes, i.e. $90\,°$ after $150\,\mathrm{ms}$, at power frequency of the voltage at the PoC. The measured current responses are presented in figure 5.23. It shows that inverter II can handle the step change while inverter I shuts down.

C. Network impedance and background voltage

In the field, typically both, a background voltage distortion and the network impedance are present. As a documented example, a large number of installed PV inverters in presence of an LV network impedance has caused severe interactions of inverters in a distribution network in the Netherlands [32].

1 Steady state

Section 5.6 has already adressed the importance and demonstrated that a reasonable background voltage distortion in presence of high network impedances (weak grids) challenges the stable operation of the inverter.

2 Transient

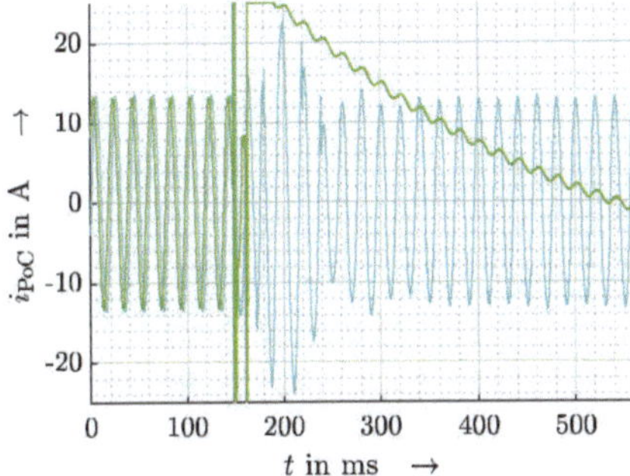

Figure 5.23: Current response of inverter I (green) and inverter II (cyan) to transients in the voltage at the PoC based on [3].

The previously individual explained changes of the background voltage and the network impedance can occur simultaneously. Generally speaking, this represents the typical case in the field. Changes of the network impedance have an impact on the background voltage distortion. One main reason is that changes in the network impedance affect other grid-connected devices [18]. Especially devices with nonlinear voltage-current characteristics cause changes in the background voltage distortion.

5.7.3 Assessment and analysis of stable operation

For white-box studies with detailed available knowledge of the entire system, i.e. also the nonlinearities of other grid-connected devices and thus a very detailed representation of the entire system, all types of instable operations can be identified and analyzed. As mentioned earlier in this work, this is not suitable and often not possible. Consequently, in future, more advanced test methods, i.e. a measurement-based framework to test for the overall stable operation, will have to include further tests besides the established tests, also in relation to the impact factors listed in table 5.9.

Component-based root causes

To development the mentioned measurement-based framework to test the overall stable operation of the inverter, the root causes, i.e. the origin of the shutdowns, related to the implemented component designs (software and hardware) has been studied based on the modular model [2]. The design of four inverters are presented in appendix A.1.5 and have been pulished in [16] while even further inverter designs have been created by the author, e.g. for chapter 5.4.

When separating the root causes into software and hardware components, the relevant software components are namely the PLL, the DC-link voltage control, the AC-current control and AI detection.

The PLL is related to the slope of the rate of change (RoC) of the frequency, e.g. a step change of the phase angle and the bandwidth of the PLL determines the maximum possible rate of change in the phase angle $\Delta\varphi$ over a given time Δt [18]. If this rate is exceeded, the PLL falls out of synchronization and tries to re-synchronize. If the rate is exceeded too largely, the PLL is not able to re-synchronize and will eventually cause the inverter to shut down. It is a challenge and a trade-off for the PLL designer to chose between a high suppression of harmonics (slow PLL, small bandwidth, good for the classical stability, bad in case of voltage transients), and the fast synchronization (fast PLL, large bandwidth, bad for classical stability but good in case of voltage transients). Consequently, the PLL designer has to chose between the sensitivity towards voltage dynamics and the performance in steady state.

As mentioned earlier, former grid synchronization algorithms have applied zero-crossing detection and failed regularily in case of double-zero crossings.

The DC-link voltage control and the AC-current control can be affected by highly distorted voltages (steady state) as well as transients at the PoC and saturate which affects the reference signals and the output signals of the control.

AI detection is specifically challenged during voltage sags and voltage interruptions though this phenomenon is related to earlier installed inverters or rather older designs which however might remain installed for another decade [18].

The main components that determine the shutdowns are overcurrent and overvoltage protection. The current and voltage ratings are chosen based on the ratings of the hardware components. Voltage sags are of relevance for the overcurrent protection and voltage swells for the overvoltage protection. Knowledge of the voltage and current ratings of the implemented overvoltage and overcurrent protection gives important information about the maximum voltage and maximum current values at the PoC for which the inverter can operate. [18]

As an advance on simulation studies, the identified overvoltage and overcurrent protection devices should be implemented in future models [19] to include the limits of a stable operation of the inverter [22] more accurately.

Immunity testing

The previous sections have demonstrated the need for more advanced immunity tests. This includes the further testing of distorted test voltages and network impedances, e.g. with and without resonances. The test voltages should include realistic multi-frequent distortions but also dynamic changes that are typical for public LV networks and relate rather to the large-signal characteristics of the inverter. In addition, the network impedance with a pronounced resonance should also be tested in combination with an applied background voltage distortion. [18]

5.8 Design considerations and recommendations

To improve the inverter design with regard to the harmonic stability and also to enhance the stable operation, the findings of this chapter can be summarized as follows:

- Impedance magnitudes and phase angles of an inverter determine the critical frequencies in combination with the measurement site for the harmonic stability [1].
- The frequency-dependent phase angles of the inverter impedances become passive with an increasing frequency due to the decreasing impact of the control (mainly determined by the bandwidth of the PLL but also of the DC-link voltage control and the AC-current control).
- The intersection of inverter impedances and network impedances relates to the decreasing magnitudes of the inverter impedance and the increasing magnitudes of the network impedances in the low frequency range (some hundred Hertz). For highly inductive network impedances with large magnitudes, this intersection occurs at low frequencies and is consequently more likely in the active frequency range of the inverter than for small network impedance magnitudes.
- Resonances in the network can enlargen the current response of the inverter compared to the respective scenario with an equivalent RL-ratio at power frequency.
- The large variety of public LV network impedances challenges the different inverters individually and relates to all previously mentioned findings in practice.
- Besides the harmonic stability based on the presence of the network impedance, the background voltage distortion including transients can also prevent a stable operation of the inverter.

As a conclusion of this chapter, the following aspects should be considered for the inverter design...
A) ...with regard to the inverter impedance characteristics (impedance shaping):

1. The inverter impedance magnitudes should be high in the low frequency range (up to some hundred Hertz) where the bandwidth of the PLL causes an active frequency region (above $-90°$) to prevent an intersection with the LV network impedance.

2. The phase angles of the inverter impedance should stay within $\pm 90°$ for higher frequencies (above some hundred Hertz) to provide a stable phase margin (passivity theory, e.g. [75]).

The inverter components and the related frequency ranges that these components affect are presented in figure 5.24 based on [30].

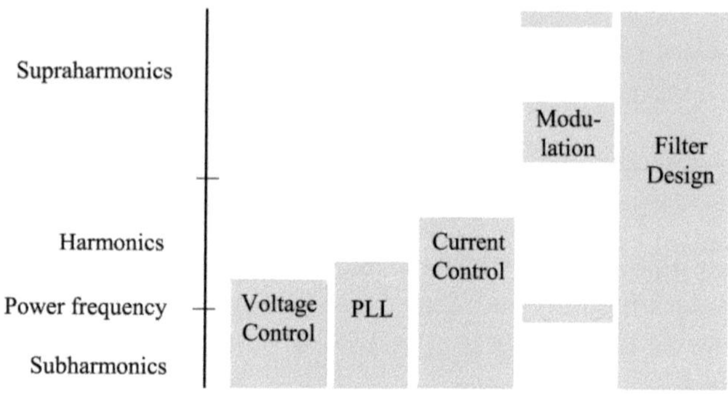

Figure 5.24: Frequency range of converter components and modulation based on [30].

B) ... with regard to the LV network representation in the analysis:

1. The representation of the LV network impedance (magnitude and phase angle) characteristics must include resonances for the harmonic stability analysis to be accurate if these resonances are present in the studied LV network.

2. In a further advanced analysis, a distortion and transients in the background voltage must be studied to conclude on the overall stable operation of the inverter.

While these findings are straightforward, the following represents a trade-off for the inverter designer in practice that requires a more holistic analysis of both, the harmonic stability of an inverter and its current response on voltage transients:
With regard to the inverter design, a narrow bandwidth of the PLL will reduce the chance of harmonic instabilities. The smaller the bandwidth of the PLL and also of the DC-link voltage control and the AC-current control, the smaller the impact of the control on the inverter impedance. Consequently, this reduces the active frequency range of the inverter impedance. However, with regard to the dynamic reaction of the inverter, the inverter will respond much slower on fast changes of $u_{\mathrm{PoC}}(t)$. High overswings in the current response as an inverter reaction on dynamic voltage changes can result in overcurrents that trigger the inverter protection, i.e. the inverter shuts down eventually. It is necessary to find a feasible trade-off between a fast reaction on voltage transients and a narrow frequency region of the inverter impedance with an active phase angle characteristics.

6 Conclusion

The conclusion of this work can be separated into firstly, a summary of the results and secondly, into future work.

6.1 Summary

The value of this work and the contribution to existing resarch can be summarized as follows: A completely black-box based harmonic stability analysis approach has been developed that is applicable on commercially available devices in the laboratory.
In detail, specific subtopics have been presented and studied:

- The state of the art has been overviewed and open research topics have been identified and related to this work.
- It has been found that the development of a consistent, purely black-box-based harmonic stability analysis required further detailed background knowledge of the inverter characteristics, e.g. a linearity analysis. This linearity analysis has been performed for commercially available single-phase inverters and is based on the analytic description of white-box representations.
- For simulation studies and the theoretic analysis, a new modular approach has been developed for white-box modeling that enables to study different implementations (hardware and software) to compare and analyze the differences and the general characteristics of commercially available devices. Hence, the model allows to generalize from the detailed study of individual single devices to an analysis of inverters in general. The general validation of the model has been achieved by comparing the frequency domain characteristics to laboratory measurements of commercially available black-box devices.
- The analysis concludes that the harmonic voltages and currents relate linearily in the harmonic frequency range but are rather time-periodic than time-invariant. A dependency on the operating power has been identified that is determined by the specific implementation not only of the hardware but also of the software, i.e. the control.

- Based on the linearity analysis of the inverter, the existing frequency sweep-based model parametrization has been improved by reducing the overall measurement duration. As an alternative, applying the principle of superposition on the exciting test voltage frequency components, the measurement duration can be reduced further. It has been found that this is essential for a holistic small-signal black-box representation of the inverter that is required for a reliable harmonic stability analysis, e.g. based on different operating powers.
- The theoretic stability analysis has been validated in the laboratory for a commercially available inverter. The validation included an adaptive test stand setup based on air coils to study different network characteristics. This test stand setup can be used without requiring expensive test stand devices and does not require the advanced background knowledge about the stability theory, thus presenting an alternative, practical but simplified approach to assess the harmonic stability of a commercially available device. By iteratively increasing the test stand inductance, the critical test stand inductance and the critical frequency region can be identified.
- The operating point dependency, i.e. the operating power, on the harmonic stability has been measured and validated. In addition, white-box simulations have shown the impact of resonances in the network impedance characteristics. Real network impedance characteristics have been used to apply the theory on the simulation models and the commercial devices probabilistically to assess the grid compatibility of the inverters in real public LV networks.
- Finally, limitations of the harmonic stability analysis and the stable operation of inverters with regard to the inverter immunity and design improvements have been discussed and exemplary measurements have been presented.

To address the research questions from the beginning in section 1.2.1, the following can be concluded:

1. *What are the input-output small-signal dependencies of inverters?*
   The small-signal model of the inverter is linear regarding the voltage-current relation at the PoC but nonlinear with regard to the operating power. Small-signal characteristics imply that the harmonic power flow is much less, i.e. below 10 % of the power at power frequency.
2. *What are suitable measurement methods to identify representations of commercially available inverter?*
   The adapted frequency sweep in 50 Hz steps with voltage amplitudes of 5 V with 0° and a suitable range of operating powers (10 % steps from 10 % to 100 %) can

be applied. More advanced and to save more time, the principle of superposition can be applied to generate multi-frequent test voltages as presented in chapter 4.3 which requires a more elaborate background knowledge and a larger effort to define the spectra of the test voltages compared to the frequency sweep.

And with regard to the main research question:
How can the harmonic stability of commercially available single-phase inverters for photovoltaic applications be assessed?
The harmonic stability can be assessed based on the inverter impedance and the network impedance. More general assessments can be performed when considering a large set of network impedances if the specific site where the inverter is installed in future is unkown. However, next to the harmonic stability, large-signal stability aspects and transients as well as the device immunity are also important for an overall stable operation of the inverter.

6.2 Future Work

Open research topics that remain as future work can be categorized into the inverter-specific research on device level and the research on system level.
On device level, the following topics have been identified as remaining future work:

- The research on device level has not presented an overall grey-box model of commercially available devices. Specifically, identification methods and a feasible representation of the inverter control is currently missing for a complete grey-box model.

- For the measurement-based assessment of the grid compatibility and a stable operation of the inverter, the immunity tests need to be adapted further, e.g. based on the presented framework in section 5.7. Feasible emission and immunity limits have to possibly be frequently adapted in the future, e.g. based on the penetration levels of PE devices, for individual (public) networks.

- For the device analysis of commercially available devices and for validation purposes, the laboratory setup has to be improved by building an adjustable test stand impedance, e.g. on the basis of the air coils, but with a T-branch including capacitors to also represent networks with resonance characteristics.

On system level, open topics include the following:

- The research on system level has to study the LTP characteristics and nonlinearities of the overall LV network in more detail. For an aggregation of all grid-connected devices and the respective interactions at the PoC (e.g. also in terms of multi-device studies), the individual linearity as presented in this work for single-phase, commercially available inverters has to be analyzed for all relevant, grid-connected devices as well as for the network equipment.

- With that detailed knowledge about the network linearities, new network representations and network descriptions, e.g. linearized time-periodic models or even nonlinear models, have to be developed and formulated.
- Measurement-based identification methods of network representations have to be improved to parametrize the new network models. These new meausrement-based identification methods have to be applied to enlargen the database of network data and to develop representative network models.

Solving the open topics on device level and on system level enables reliable harmonic black-box studies of LV networks with a very high penetration rate of PE devices, e.g. future public LV networks.

References

[1] Elias Kaufhold, Jan Meyer, Sascha Müller, and Peter Schegner. Probabilistic Stability Analysis for Commercial Low Power Inverters Based on Measured Grid Impedances. In *2019 9th International Conference on Power and Energy Systems (ICPES)*, pages 1–6, Perth, dec 2019. IEEE. doi:10.1109/ICPES47639.2019.9105647.

[2] Elias Kaufhold, Jan Meyer, and Peter Schegner. Modular White-Box Model of single-phase Photovoltaic Systems for Harmonic Studies. In *2019 IEEE Milan PowerTech*, pages 1–6, Milano, jun 2019. IEEE. doi:10.1109/PTC.2019.8810426.

[3] Elias Kaufhold, Jan Meyer, and Peter Schegner. Measurement Framework for Analysis of Dynamic Behavior of Single-Phase Power Electronic Devices. *Renewable Energy and Power Quality Journal*, 18:494–499, jun 2020. doi:10.24084/repqj18.407.

[4] Elias Kaufhold, Jan Meyer, and Peter Schegner. Transient response of single phase photovoltaic inverters to step changes in supply voltage distortion. In *2020 19th International Conference on Harmonics and Quality of Power (ICHQP)*, pages 1–6, Dubai, jul 2020. IEEE. doi:10.1109/ICHQP46026.2020.9177918.

[5] Elias Kaufhold, Jan Meyer, Peter Schegner, Ahmed S. Abdelsamad, and Johanna M.A. Myrzik. Comparison of solvers for time-domain simulations of single-phase photovoltaic systems. In *2020 International Conference on Smart Grids and Energy Systems (SGES)*, pages 550–555, Perth, nov 2020. IEEE. doi:10.1109/SGES51519.2020.00103.

[6] Elias Kaufhold, Jan Meyer, and Peter Schegner. Impact of grid impedance and their resonance on the stability of single-phase PV-inverters in low voltage grids. In *2020 IEEE 29th International Symposium on Industrial Electronics (ISIE)*, pages 880–885, Delft, jun 2020. IEEE. doi:10.1109/ISIE45063.2020.9152571.

[7] Elias Kaufhold, Simon Grandl, Jan Meyer, and Peter Schegner. Feasibility of Black-Box Time Domain Modeling of Single-Phase Photovoltaic Inverters Using Artificial Neural Networks. *Energies*, 14(8):2118, apr 2021. doi:10.3390/en14082118.

[8] Elias Kaufhold, Jan Meyer, and Peter Schegner. Black-box identification of grid-side filter circuit for improved modelling of single-phase power electronic devices for harmonic studies. *Electric Power Systems Research*, 199:107421, oct 2021. doi:10.1016/j.epsr.2021.107421.

[9] Elias Kaufhold, Jan Meyer, and Peter Schegner. Measurement-based identification of DC-link capacitance of single-phase power electronic devices for grey-box modelling. *IEEE Transactions on Power Electronics*, pages 1–1, 2021. doi:10.1109/TPEL.2021.3127241.

[10] Elias Kaufhold, Jan Meyer, Carlos Duque, and Peter Schegner. Simplified method for measurement-based black-box harmonic stability assessment of single-phase power electronic devices. In *2021 IEEE 11th International Workshop on Applied Measurements for Power Systems (AMPS)*, pages 1–6. IEEE, sep 2021. doi:10.1109/AMPS50177.2021.9586005.

[11] Elias Kaufhold, Jan Meyer, and Peter Schegner. Fast measurement-based identification of small signal behaviour of commercial single-phase inverters. In *2021 IEEE 11th International Workshop on Applied Measurements for Power Systems (AMPS)*, pages 1–6. IEEE, sep 2021. doi:10.1109/AMPS50177.2021.9586029.

[12] Elias Kaufhold, Jan Meyer, and Peter Schegner. Impact of harmonic distortion on the supraharmonic emission of pulsewidth modulated single-phase power electronic devices. *Renewable Energy and Power Quality Journal*, 19:577–582, sep 2021. doi:10.24084/repqj19.352.

[13] Elias Kaufhold, Jan Meyer, and Peter Schegner. Research needs for harmonic stability analysis of power electronic devices in low voltage networks. In *IECON 2021 – 47th Annual Conference of the IEEE Industrial Electronics Society*, pages 1–6. IEEE, oct 2021. doi:10.1109/IECON48115.2021.9589680.

[14] Elias Kaufhold, Jan Meyer, and Peter Schegner. Measurement-Based Black-Box Harmonic Stability Analysis of Commercial Single-Phase Inverters in Public Low-Voltage Networks. *Solar*, 2(1):64–80, mar 2022. doi:10.3390/solar2010005.

[15] Elias Kaufhold, Carlos A. Duque, Jan Meyer, and Peter Schegner. Measurement-Based Black-Box Harmonic Stability Assessment of Single-Phase Power Electronic Devices Based on Air Coils. *IEEE Transactions on Instrumentation and Measurement*, 71:1–9, 2022. doi:10.1109/TIM.2022.3160556.

[16] Elias Kaufhold, Jan Meyer, and Peter Schegner. A modular time-domain model of single-phase photovoltaic inverters enabling realistic harmonic large-scale sim-

ulations in low voltage networks. In *2022 IEEE 20th International Power Electronics and Motion Control Conference (PEMC)*, pages 481–486. IEEE, sep 2022. doi:10.1109/PEMC51159.2022.9962930.

[17] Elias Kaufhold, Sascha Müller, Jan Meyer, Johanna Myrzik, and Peter Schegner. Impact of DC power on the small signal characteristics of single-phase photovoltaic inverters in frequency domain. In *2023 International Conference on Clean Electrical Power (ICCEP)*, pages 526–531, Terrasini, jun 2023. IEEE. doi:10.1109/ICCEP57914.2023.10247364.

[18] Elias Kaufhold, J. Meyer, J. Myrzik, and P. Schegner. Framework to assess the stable operation of commercially available single-phase inverters for photovoltaic applications in public low voltage networks. *Renewable Energy and Power Quality Journal*, 2023.

[19] Elias Kaufhold, Jan Meyer, Johanna Myrzik, and Peter Schegner. Harmonic Stability Assessment of Commercially Available Single-Phase Photovoltaic Inverters Considering Operating-Point Dependencies. *Solar*, 3(3):473–486, aug 2023. doi:10.3390/solar3030026.

[20] Elias Kaufhold, Sascha Müller, Jan Meyer, Johanna Myrzik, and Peter Schegner. Scalability of Harmonic Emission Generated by Single-phase Photovoltaic Inverters. In *PEDG 2023 - 2023 IEEE 14th International Symposium on Power Electronics for Distributed Generation Systems*, pages 678–683. IEEE, jun 2023. doi:10.1109/PEDG56097.2023.10215231.

[21] Elias Kaufhold, Jan Meyer, and Peter Schegner. Harmonic stability assessment of commercially available single-phase photovoltaic inverters in public low voltage networks. In *1st Helgoland Power and Energy Conference (HPEC 2023)*, Helgoland, 2023. doi:10.15488/15895.

[22] Elias Kaufhold, Jan Meyer, Johanna Myrzik, and Peter Schegner. Limits of Harmonic Stability Analysis for Commercially Available Single-Phase Inverters for Photovoltaic Applications. *Solar*, 4(3):387–400, jul 2024. doi:10.3390/solar4030017.

[23] Elias Kaufhold and Jan Meyer. Chapter 23: A measurement framework for the analysis of the dynamic behaviour of single-phase power electronic devices. In *Trends in Renewable Energy and Power Quality*, pages 359–375. Cambridge Scholars Publishing, 2024.

[24] Ahmed S Abdelsamad, M.A. Johanna Myrzik, Elias Kaufhold, Jan Meyer, and Peter Schegner. Nonlinear identification approach for black-box modeling of voltage

source converter harmonic characteristics. In *2020 IEEE Electric Power and Energy Conference (EPEC)*, pages 1–5, Edmonton, AB, nov 2020. IEEE. doi:10.1109/EPEC48502.2020.9320079.

[25] Ahmed S. Abdelsamad, Johanna M. A. Myrzik, Elias Kaufhold, Jan Meyer, and Peter Schegner. Voltage-Source Converter Harmonic Characteristic Modeling Using Hammerstein-Wiener Approach Modélisation des caractéristiques harmoniques d'un convertisseur tension-source à l'aide de l'approche Hammerstein-Wiener. *IEEE Canadian Journal of Electrical and Computer Engineering*, pages 1–9, 2021. doi:10.1109/ICJECE.2021.3088325.

[26] Zafar Iqbal, Sasa Z. Djokic, Jan Meyer, Sascha Müller, and Elias Kaufhold. Filter and Controller Identification for Stability Analysis of a Grid-Connected 3-Phase PV Inverter. In *2022 20th International Conference on Harmonics and Quality of Power (ICHQP)*, pages 1–6. IEEE, may 2022. doi:10.1109/ICHQP53011.2022.9808533.

[27] Ahmed S. Abdelsamad, Johanna M. A. Myrzik, Elias Kaufhold, Jan Meyer, and Peter Schegner. Comparison of harmonic models for a commercial battery energy storage system in charging and discharging mode. In *2022 20th International Conference on Harmonics and Quality of Power (ICHQP)*, pages 1–6. IEEE, may 2022. doi:10.1109/ICHQP53011.2022.9808696.

[28] REN21. Renewables 2022 Global status report, 2022.

[29] Dejan Matvoz and Matjaz Miklavcic. A practical case of harmonic current issues operating a small rooftop pv plant. In *25th International Conference on Electricity Distribution*, Madrid, 2019.

[30] Xiongfei Wang and Frede Blaabjerg. Harmonic Stability in Power Electronic-Based Power Systems: Concept, Modeling, and Analysis. *IEEE Transactions on Smart Grid*, 10(3):2858–2870, may 2019. doi:10.1109/TSG.2018.2812712.

[31] Erik Möllerstedt and Bo Bernhardsson. Out of control because of harmonics-an analysis of the harmonic response of an inverter locomotive. *IEEE Control Systems*, 20(4):70–81, aug 2000. doi:10.1109/37.856180.

[32] Johan. H. R. Enslin and Peter J. M. Heskes. Harmonic Interaction Between a Large Number of Distributed Power Inverters and the Distribution Network. *IEEE Transactions on Power Electronics*, 19(6):1586–1593, nov 2004. doi:10.1109/TPEL.2004.836615.

[33] Matthias Brendel and Gerald Traufetter. Knall auf hoher See. *Spiegel*, 2014.

[34] Michael Höckel, Andreas Gut, Michel Arnal, Roman Schild, Peter Steinmann, and Stefan Schori. Measurement of voltage instabilities caused by inverters in weak grids. *CIRED - Open Access Proceedings Journal*, 2017(1):770–774, oct 2017. doi:10.1049/oap-cired.2017.0997.

[35] Moayed Moghbel, Simon Glenister, Martina Calais, Farhad Shahnia, Craig Carter, David Edwards, David Stephens, Luke Jones, and Pierce Trinkl. Fluctuations in the Output Power of Photovoltaic Systems Distributed Across a Town with an Isolated Power System Using High-Resolution Data. In *2019 9th International Conference on Power and Energy Systems (ICPES)*, pages 1–5. IEEE, dec 2019. doi:10.1109/ICPES47639.2019.9105539.

[36] Remus Teodorescu, Marco Liserre, and Pedro Rodríguez. *Grid Converters for Photovoltaic and Wind Power Systems.* John Wiley & Sons, Ltd, Chichester, UK, jan 2011. doi:10.1002/9780470667057.

[37] Stefan Reichert, Gerd Griepentrog, and Benjamin Stickan. Comparison between grid-feeding and grid-supporting inverters regarding power quality. In *2017 IEEE 8th International Symposium on Power Electronics for Distributed Generation Systems (PEDG)*, pages 1–4. IEEE, apr 2017. doi:10.1109/PEDG.2017.7972536.

[38] CENELEC. EN 50160: Voltage characteristics of electricity supplied by public electricity networks, 2022.

[39] VDE-FNN. Power Generating Plants in the Low Voltage Network (VDE-AR-N 4105). Technical report, 2018.

[40] Donald Grahame Holmes and Thomas A. Lipo. *Pulse Width Modulation for Power Converters.* IEEE, 2003. doi:10.1109/9780470546284.

[41] Steven R. Hall and Norman M. Wereley. Generalized Nyquist Stability Criterion for Linear Time Periodic Systems. In *1990 American Control Conference*, pages 1518–1525, 1990.

[42] Atle Rygg, Marta Molinas, Chen Zhang, and Xu Cai. Coupled and decoupled impedance models compared in power electronics systems. Technical Report 1, oct 2016. arXiv:1610.04988.

[43] Junbum Kwon, Xiongfei Wang, Claus Leth Bak, and Frede Blaabjerg. Harmonic instability analysis of single-phase grid connected converter using Harmonic State Space (HSS) modeling method. In *2015 IEEE Energy Conversion Congress and Exposition (ECCE)*, pages 2421–2428. IEEE, sep 2015. doi:10.1109/ECCE.2015.7310001.

[44] Rudolf E. Kalman. A New Approach to Linear Filtering and Prediction Problems. *Journal of Basic Engineering*, 82(1):35–45, mar 1960. doi:10.1115/1.3662552.

[45] Bing Hao Lin, Jin Tong Tsai, and Kuo Lung Lian. A Non-Invasive Method for Estimating Circuit and Control Parameters of Voltage Source Converters. *IEEE Transactions on Circuits and Systems I: Regular Papers*, 66(12):4911–4921, dec 2019. doi:10.1109/TCSI.2019.2929783.

[46] Nopporn Patcharaprakiti, Krissanapong Kirtikara, Veerapol Monyakul, Dhirayut Chenvidhya, Jatturit Thongpron, Anawach Sangswang, and Ballang Muenpinij. Modeling of single phase inverter of photovoltaic system using Hammerstein–Wiener nonlinear system identification. *Current Applied Physics*, 10(3):S532–S536, may 2010. doi:10.1016/j.cap.2010.02.025.

[47] Laura M. Wiesermann. *Developing a transient photovoltaic inverter model in opendss using the Hammerstein-Wiener mathematical Structure*. PhD thesis, University of Pittsburgh, 2017.

[48] Maurizio Fauri. Harmonic modelling of non-linear load by means of crossed frequency admittance matrix. *IEEE Transactions on Power Systems*, 12(4):1632–1638, 1997. doi:10.1109/59.627869.

[49] Yuanyuan Sun, Guibin Zhang, Wilsun Xu, and Julio G. Mayordomo. A Harmonically Coupled Admittance Matrix Model for AC/DC Converters. *IEEE Transactions on Power Systems*, 22(4):1574–1582, nov 2007. doi:10.1109/TPWRS.2007.907514.

[50] P. Friedmann, C. E. Hammond, and Tze-Hsin Woo. Efficient numerical treatment of periodic systems with application to stability problems. *International Journal for Numerical Methods in Engineering*, 11(7):1117–1136, jan 1977. URL: https://onlinelibrary.wiley.com/doi/10.1002/nme.1620110708, doi:10.1002/nme.1620110708.

[51] Sjef Cobben, Wil Kling, and Johanna Myrzik. The Making and Purpose of Harmonic Fingerprints. In *19th International Conference on Electricity Distribution*, number 0764, pages 21–24, 2007.

[52] Daniele Gallo, Roberto Langella, Mario Luiso, Alfredo Testa, and Neville R. Watson. A New Test Procedure to Measure Power Electronic Devices' Frequency Coupling Admittance. *IEEE Transactions on Instrumentation and Measurement*, 67(10):2401–2409, oct 2018. doi:10.1109/TIM.2018.2819318.

[53] Roberto Langella, Alfredo Testa, Jan Meyer, Friedemann Moller, Robert Stiegler, and Sasa Z. Djokic. Experimental-Based Evaluation of PV Inverter Harmonic and

Interharmonic Distortion Due to Different Operating Conditions. *IEEE Transactions on Instrumentation and Measurement*, 65(10):2221–2233, oct 2016. doi:10.1109/TIM.2016.2554378.

[54] X. Xu, A. Collin, S. Djokic, S. Müller, J. Meyer, J. Myrzik, R. Langella, and A. Testa. Aggregate harmonic fingerprint models of PV inverters. part 1: Operation at different powers. In *2018 18th International Conference on Harmonics and Quality of Power (ICHQP)*, pages 1–6. IEEE, may 2018. doi:10.1109/ICHQP.2018.8378824.

[55] Roberto Langella, Alfredo Testa, Joaquin E. Caicedo, Andres A. Romero, Humberto C. Zini, Jan Meyer, and Neville R. Watson. On the use of fourier descriptors for the assessment of frequency coupling matrices of power electronic devices. In *2018 18th International Conference on Harmonics and Quality of Power (ICHQP)*, volume 2018-May, pages 1–6. IEEE, may 2018. doi:10.1109/ICHQP.2018.8378908.

[56] N. Patcharaprakiti, K. Kirtikara, D. Chenvidhya, V. Monyakul, and B. Muenpinij. Modeling of Single Phase Inverter of Photovoltaic System Using System Identification. In *2010 Second International Conference on Computer and Network Technology*, pages 462–466. IEEE, 2010. doi:10.1109/ICCNT.2010.120.

[57] Prabha Kundur. Power System Stability and Control, 1994.

[58] Nikos Hatziargyriou, Jovica Milanovic, Claudia Rahmann, Venkataramana Ajjarapu, Claudio Canizares, Istvan Erlich, David Hill, Ian Hiskens, Innocent Kamwa, Bikash Pal, Pouyan Pourbeik, Juan Sanchez-Gasca, Aleksandar Stankovic, Thierry Van Cutsem, Vijay Vittal, and Costas Vournas. Definition and Classification of Power System Stability – Revisited and Extended. *IEEE Transactions on Power Systems*, 36(4):3271–3281, jul 2021. doi:10.1109/TPWRS.2020.3041774.

[59] IEC. IEC 61000-3-2: Electromagnetic compatibility (EMC) - Part 3-2: Limits - Limits for harmonic current emissions (equipment input current $\leq$ 16 A per phase), 2018.

[60] Valerio Salis, Alessandro Costabeber, Stephen M. Cox, Francesco Tardelli, and Pericle Zanchetta. Experimental Validation of Harmonic Impedance Measurement and LTP Nyquist Criterion for Stability Analysis in Power Converter Networks. *IEEE Transactions on Power Electronics*, 34(8):7972–7982, aug 2019. doi:10.1109/TPEL.2018.2880935.

[61] Jian Sun. Impedance-based stability criterion for grid-connected inverters. *IEEE Transactions on Power Electronics*, 26(11):3075–3078, nov 2011. doi:10.1109/TPEL.2011.2136439.

[62] D. Gabor. Theory of communication. Part 1: The analysis of information. *Journal of the Institution of Electrical Engineers - Part III: Radio and Communication Engineering*, 93(26):429–441, nov 1946. doi:10.1049/ji-3-2.1946.0074.

[63] Cong Wang, Andrey Bernstein, Jean-Yves Le Boudec, and Mario Paolone. Explicit Conditions on Existence and Uniqueness of Load-Flow Solutions in Distribution Networks. *IEEE Transactions on Smart Grid*, 9(2):953–962, mar 2018. doi:10.1109/TSG.2016.2572060.

[64] Andreas Martin Kettner and Mario Paolone. On the Properties of the Compound Nodal Admittance Matrix of Polyphase Power Systems. *IEEE Transactions on Power Systems*, 34(1):444–453, jan 2019. doi:10.1109/TPWRS.2018.2863671.

[65] Juan Luis Agorreta, Mikel Borrega, Jesús López, and Luis Marroyo. Modeling and control of N-paralleled grid-connected inverters with LCL filter coupled due to grid impedance in PV plants. *IEEE Transactions on Power Electronics*, 26(3):770–785, mar 2011. doi:10.1109/TPEL.2010.2095429.

[66] T. Bussato, M. Bollen, and S. Rönnberg. Photovoltaics and Harmonics in Low-Voltage Networks, 2018.

[67] Vázquez-Guzmán Gerardo, Martínez-Rodríguez Pánfilo Raymundo, and Sosa-Zúñiga José Miguel. High Efficiency Single-Phase Transformer-less Inverter for Photovoltaic Applications, apr 2015. doi:10.1016/j.riit.2015.03.002.

[68] Matthias Klatt, Robert Stiegler, Jan Meyer, and Peter Schegner. Generic frequency-domain model for the emission of PWM-based power converters in the frequency range from 2 to 150 kHz. *IET Generation, Transmission & Distribution*, 13(24):5478–5486, 2019. doi:10.1049/iet-gtd.2018.6985.

[69] R. Stiegler, J. Meyer, P. Schegner, and D. Chakravorty. Measurement of network harmonic impedance in presence of electronic equipment. In *2015 IEEE International Workshop on Applied Measurements for Power Systems (AMPS)*, pages 1–6. IEEE, sep 2015. doi:10.1109/AMPS.2015.7312737.

[70] Robert Stiegler, Jan Meyer, Stefan Schori, Michael Höckel, Jirí Drápela, and Tomász Hanzlík. Survey of network impedance in the frequency range 2-9 kHZ in public low voltage networks in AT / CH / CZ / GE. In *25th International Conference on Electricity Distribution*, number June, pages 3–6, 2019.

[71] Elaine Santos, Mahdi Khosravy, Marcelo A.A. Lima, Augusto S. Cerqueira, and Carlos A. Duque. ESPRIT associated with filter bank for power-line harmonics,

sub-harmonics and inter-harmonics parameters estimation. *International Journal of Electrical Power and Energy Systems Energy Systems*, 118:105731, jun 2020. doi:10.1016/j.ijepes.2019.105731.

[72] Tatiano Busatto, Anders Larsson, Sarah K. Ronnberg, and Math H. J. Bollen. Including Uncertainties From Customer Connections in Calculating Low-Voltage Harmonic Impedance. *IEEE Transactions on Power Delivery*, 34(2):606–615, apr 2019. doi:10.1109/TPWRD.2018.2881222.

[73] IEC. IEC 61000-3-12: Electromagnetic compatibility (EMC) - Part 3-12: Limits - Limits for harmonic currents produced by equipment connected to public low-voltage systems with input current >16 A and ≤ 75 A per phase, 2011.

[74] IEC. IEC 61000-4-13: Electromagnetic compatibility (EMC) - Part 4-13: Testing and measurement techniques - Harmonics and interharmonics including mains signalling at a.c. power port, low frequency immunity tests. Technical report, 2002.

[75] Lennart Harnefors, Xiongfei Wang, Alejandro G Yepes, and Frede Blaabjerg. Passivity-Based Stability Assessment of Grid-Connected VSCs—An Overview. *IEEE Journal of Emerging and Selected Topics in Power Electronics*, 4(1):116–125, mar 2016. doi:10.1109/JESTPE.2015.2490549.

[76] Ahmad Ale Ahmad, Student Member, Adib Abrishamifar, and Mohammad Farzi. A New Design Procedure for Output LC Filter of Single Phase Inverters. In *2010 3rd International Conference on Power Electronics and Intelligent Transportation System*, number 1, pages 86–91, 2010.

[77] Jozef Sedo and Slavomir Kascak. Design of output LCL filter and control of single-phase inverter for grid-connected system, dec 2017. doi:10.1007/s00202-017-0617-0.

[78] Aleksandr Reznik, Marcelo Godoy Simoes, Ahmed Al-Durra, and S M Muyeen. LCL Filter design and performance analysis for grid-interconnected systems. *IEEE Transactions on Industry Applications*, 50(2):1225–1232, mar 2014. doi:10.1109/TIA.2013.2274612.

[79] Remus Narcis Beres, Xiongfei Wang, Marco Liserre, Frede Blaabjerg, and Claus Leth Bak. A Review of Passive Power Filters for Three-Phase Grid-Connected Voltage-Source Converters. *IEEE Journal of Emerging and Selected Topics in Power Electronics*, 4(1):54–69, mar 2016. doi:10.1109/JESTPE.2015.2507203.

[80] Xinbo Ruan, Xuehua Wang, Donghua Pan, Dongsheng Yang, Weiwei Li, and Chenlei Bao. Control techniques for LCL-type grid-connected inverters, jul 2018. doi:10.1007/978-981-10-4277-5.

[81] MathWorks. Masking Fundamentals, 2018. URL: https://www.mathworks.com/help/simulink/ug/block-masks.html. accessed 2024/01/21.

[82] Simone Buso and Paolo Mattavelli. Digital control in power electronics. Technical Report 1, jan 2006. doi:10.2200/S00047ED1V01Y200609PEL002.

[83] M. Ciobotaru, T. Kerekes, R. Teodorescu, and A. Bouscayrol. PV inverter simulation using MATLAB/Simulink graphical environment and PLECS blockset. In *IECON 2006 - 32nd Annual Conference on IEEE Industrial Electronics*, pages 5313–5318. IEEE, nov 2006. doi:10.1109/IECON.2006.347663.

[84] IEC. IEC 60725: Consideration of reference impedances and public supply network impedances for use in determining the disturbance characteristics of electrical equipment having a rated current below 75 A per phase, 2012.

[85] Spitzenberger & Spies. Photovoltaic simulators PVS, PVS/HV and PVS/LV series, 2021. URL: https://spitzenberger.de/download.ashx?Weblink=1015. accessed 2024/01/21.

[86] Bourns. 2049 Series Medium Duty 2-Electrode Gas Discharge Tube, 2019. URL: https://www.bourns.com/docs/product-datasheets/2049.pdf. accessed 2024/01/21.

[87] Dewetron. SE-CUR-CLAMP-150-DC, 2024. URL: https://ccc.dewetron.com/dl/5bd05307-d80c-4e42-b030-7f89d9c49a3c. accessed 2024/01/21.

[88] Dewetron. HSI-LV Isolated low voltage module, 2024. URL: https://ccc.dewetron.com/dl/53970d9a-%0A05e0-44f1-b968-0709d9c49861. accessed 2024/01/21.

[89] Dewetron. HSI-HV Isolated high voltage module, 2024. URL: http://ccc.dewetron.com/dl/53970d9a-fc10-44f0-a97f-0709d9c49861. accessed 2024/01/21.

[90] Dewetron. Power Measurement. URL: http://www.dewesolutions.sg/uploads/7/7/1/6/7716986/dewetron-apps_power_pm_de_b090315e.pdf. accessed 2024/01/21.

[91] Omicron. CMC 256plus. URL: https://www.omicronenergy.com/download/document/8712CBB4-7F53-482B-A3D1-CC434574BE10/. accessed 2024/01/21.

[92] D. Roy Choudhury. *Networks and Systems.* New Age International, 1988.

[93] Mouser Electronics. Passive Components, 2020. URL: https://eu.mouser.com/c/passive-components/. accessed 2024/01/21.

[94] Hammam Soliman, Huai Wang, and Frede Blaabjerg. A Review of the Condition Monitoring of Capacitors in Power Electronic Converters. *IEEE Transactions on Industry Applications*, 52(6):4976–4989, 2016. doi:10.1109/TIA.2016.2591906.

[95] Prasanth Sundararajan, Mohamed Halick Mohamed Sathik, Firman Sasongko, Chuan Seng Tan, Josep Pou, Frede Blaabjerg, and Amit Kumar Gupta. Condition Monitoring of DC-Link Capacitors Using Goertzel Algorithm for Failure Precursor Parameter and Temperature Estimation. *IEEE Transactions on Power Electronics*, 35(6):6386–6396, jun 2020. doi:10.1109/TPEL.2019.2951859.

[96] Md Waseem Ahmad, P. Nandha Kumar, Abhinav Arya, and Sandeep Anand. Noninvasive technique for DC-link capacitance estimation in single-phase inverters. *IEEE Transactions on Power Electronics*, 33(5):3693–3696, 2018. doi:10.1109/TPEL.2017.2762341.

[97] Yu Wu and Xiong Du. A VEN Condition Monitoring Method of DC-Link Capacitors for Power Converters. *IEEE Transactions on Industrial Electronics*, 66(2):1296–1306, 2019. doi:10.1109/TIE.2018.2835393.

[98] Kai Yao, Weijie Tang, Xiaopeng Bi, and Jianguo Lyu. An Online Monitoring Scheme of DC-Link Capacitor's ESR and C for a Boost PFC Converter. *IEEE Transactions on Power Electronics*, 31(8):5944–5951, aug 2016. doi:10.1109/TPEL.2015.2496267.

[99] Yi Liu, Huai Wang, Meng Huang, Xiaoming Zha, and Yushuang Liu. Influence of DC Link Capacitance on Power Efficiency of Single-Phase Inverter. In *2018 IEEE Energy Conversion Congress and Exposition, ECCE 2018*, pages 1498–1504, 2018. doi:10.1109/ECCE.2018.8557744.

[100] Junbum Kwon, Xiongfei Wang, Frede Blaabjerg, and Claus Leth Bak. Comparison of LTI and LTP models for stability analysis of grid converters. In *2016 IEEE 17th Workshop on Control and Modeling for Power Electronics (COMPEL)*, pages 1–8. IEEE, jun 2016. doi:10.1109/COMPEL.2016.7556769.

[101] Atle Rygg, Marta Molinas, Chen Zhang, and Xu Cai. A Modified Sequence-Domain Impedance Definition and Its Equivalence to the dq-Domain Impedance

Definition for the Stability Analysis of AC Power Electronic Systems. *IEEE Journal of Emerging and Selected Topics in Power Electronics*, 4(4):1383–1396, 2016. arXiv:1605.00526, doi:10.1109/JESTPE.2016.2588733.

[102] Gaston Floquet. Sur les équations différentielles linéaires à coefficients périodiques, 1883.

[103] G. W. Hill. On the part of the motion of the lunar perigee which is a function of the mean motions of the sun and moon, 1886. doi:10.1007/BF02417081.

[104] Émile Leonard Mathieu. Course de physique mathématique. Technical report, Paris, 1873.

[105] Vladimir Cuk. Low-frequency distortion in distribution networks, 2013. doi:10.6100/IR759351.

[106] C. A. Desoer and Y. T. Wang. On the generalized Nyquist stability criterion. In *1979 18th IEEE Conference on Decision and Control including the Symposium on Adaptive Processes*, pages 580–586. IEEE, dec 1979. doi:10.1109/CDC.1979.270248.

[107] Valerio Salis, Alessandro Costabeber, Stephen M. Cox, and Pericle Zanchetta. Stability Assessment of Power-Converter-Based AC systems by LTP Theory: Eigenvalue Analysis and Harmonic Impedance Estimation. *IEEE Journal of Emerging and Selected Topics in Power Electronics*, 5(4):1513–1525, dec 2017. doi:10.1109/JESTPE.2017.2714026.

[108] C. F. M. Almeida and N. Kagan. Harmonic coupled norton equivalent model for modeling harmonic-producing loads. In *Proceedings of 14th International Conference on Harmonics and Quality of Power - ICHQP 2010*, pages 1–9. IEEE, sep 2010. doi:10.1109/ICHQP.2010.5625491.

List of Figures

List of Tables

Appendix

Appendix A

Simulation model

A.1 Modular model

For the implementation of the simulation model, a modular modeling approach has been developed by the author as published in [2] and is described in the following of this section A.1. The similarities of the simulation model and commercially available devices have been validated to ensure the suitability of the model and been published in [16].
It is a challenge for large-scale studies to design and implement a larger number of different inverter designs. Typically, manufacturers improve the inverter designs based on previous designs. However, the design can vary largely between manufacturers. In literature and when studying commercially available devices, the specific parameters and the general topology in terms of hardware components and software algorithms are not disclosed. This presents a challenge for academic studies of different designs, e.g. of different manufacturers, in general.
The main idea of the modular modeling approach is to design different inverter models by simply exchanging individual device components to create new makes of inverters. Since only parts of the inverter have to be adapted, only a reduced design and implementation effort is needed compared to creating entirely new device designs. The software component parameters and the hardware component parameters are caluclated automatically based on design formulas and a set of required and pre-defined model parameters and initial properties, e.g. the DC-link voltage ripple. This automatic calculation requires the interchangeability of all inverter components and thus requires a generic model structure as presented in figure 2.2 and figure 2.3.
Furthermore, the modular modeling approach allows a component-specific analysis and the comparison of different implementations. By systematically exchanging design components, the impact of the changed design components on the grid-side current and the harmonic stability of the inverter can be assessed. In addition, the number of designed inverters can be expanded with time by using the previous implementations and developing a suitable set of new component designs.

The design of the modular model in this work is oriented on a rated power of 3.9 kW. For simulations, the rated power is rather a guideline and is used to choose suitable values for the component parameters with regard to the expected operating power since overcurrent and overvoltage protection devices are not implemented in the model.

A.1.1 Hardware components

AC-side filter circuit

For the AC-side filter circuit, different topologies are possible, such as L-, LC-, LCL- and cascaded topologies, e.g. $LCLC$-filters. However, the L-filter is typically not used in practice for single-phase inverters since it requires large inductance values and consequently large and heavy coils that increase the cost and the weight of the inverter. For academic purposes, an $L-$filter design consisting only of the filter inductance L_{f} and the respective resistance $R_{\mathrm{f}\,L}$ is still available as one filter option but the two different LCL-filters have been used for the presented studies due to their practical relevance. The three sets of filter parameters are listed in table A.1 according to figure A.1. The main model components of an LCL-filter circuit are the device-side inductor $L_{\mathrm{f\,d}}$, the grid-side inductance $L_{\mathrm{f\,g}}$ and the T-branch capacity C_{f}. The losses in the coils are represented by the parasitic resistances $R_{\mathrm{f\,d}}$ and the resistance $R_{\mathrm{f\,g}}$. A damping resistor with the resistance $R_{\mathrm{f}\,C}$ is typically implemented in series to the capacitance to shape the frequency response of the overall filter circuit. The voltages and currents have been named in relation to the filter identification method that is introduced in the later sections D.1. For the calculation of

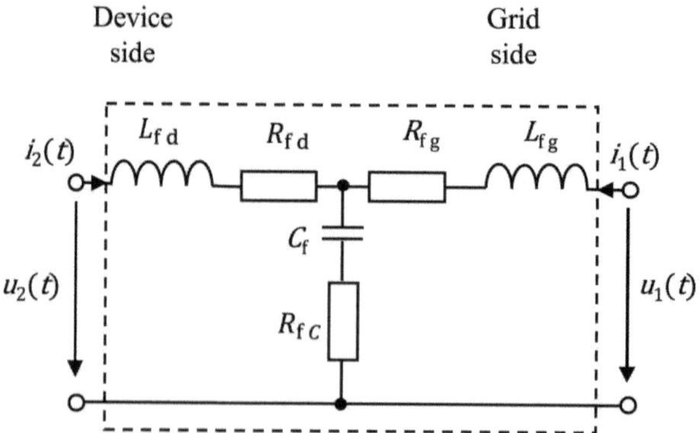

Figure A.1: Scheme of an LCL-filter.

the parameters of the LCL-filter I, [76–80] have been considered. The inductance and capacitance values of the LCL-filter II have been proposed in [36].

Table A.1: AC-side filter circuit parameters [2].

L-filter		*LCL*-filter I		*LCL*-filter II [36]	
Parameter	Value	Parameter	Value	Parameter	Value
L_{f}	10 mH	$R_{\mathrm{f}\,C}$	2.465 Ω	$R_{\mathrm{f}\,C}$	8.2 Ω
$R_{\mathrm{f}\,L}$	30 mΩ	C_{f}	7 μF	C_{f}	5 μF
		$L_{\mathrm{f\,g}}$	0.476 mH	$L_{\mathrm{f\,g}}$	2 mH
		$R_{\mathrm{f\,g}}$	20 mΩ	$R_{\mathrm{f\,g}}$	8 mΩ
		$L_{\mathrm{f\,d}}$	2 mH	$L_{\mathrm{f\,d}}$	4 mH
		$R_{\mathrm{f\,d}}$	20 mΩ	$R_{\mathrm{f\,d}}$	8 mΩ

DC-link capacitance

The aggregated capacitors of the capacitor bank are represented by the DC-link capacitance C_{DC} while resistive elements are neglected.

Switches

The switches are implemented as IGBTs with a switching frequency, the carrier frequency f_{car}, of 8 kHz and the switching time T_{sw} for

$$T_{\mathrm{sw}} = \frac{1}{f_{\mathrm{car}}}. \tag{A.1}$$

Based on the unipolar modulation (see appendix A.1.2), this leads to an effective switching frequency of 16 kHz.

A.1.2 Software components

Phase locked loop

For state of the art PV inverters, the grid synchronization algorithm is a PLL. Classical PLLs are related to either a dq-frame or an $\alpha\beta$-frame. Both frames are based on orthogonal signals, e.g. the dq-components or the $\alpha\beta$-components, if the 0-component is neglected, e.g. when detecting the amplitudes and the phase angles of the input signals. When studying single-phase devices compared to three-phase devices, there is typically only a one-dimensional input signal, i.e. the line-neutral voltage at AC-side voltage. Consequently, a second signal component has to be created, e.g. an artificial β- or q-component. This component is shifted by 90 ° compared to the original signal. In practice, there are different ways to create this second component, The algorithm that creates the second component is called the OSG. For the OSG, different implementations are possible. Many PLLs are based on filter techniques and transformations. The simplest OSG is a time delay (TD) that assumes a power frequency of the voltage, e.g. a respective feed-forward angular frequency ω_{ff} of $2\pi 50\,\mathrm{Hz}$, and delays the input signal by a fixed time, i.e. one fourth of the period length, which is why it is also called the T4D PLL. A visualization of the implemented T4D PLL is shown in figure A.2. To reset the detected phase angle of the voltage at the PoC after 360° to zero, the Euclidean division (REM) function is used.

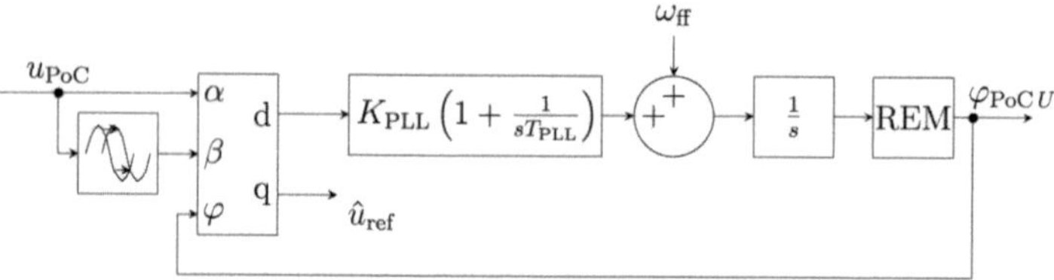

Figure A.2: Modell of a T4D PLL [2].

A PLL based on the Park transform is the IPT PLL that is implemented in the model as well and presented in figure A.3.

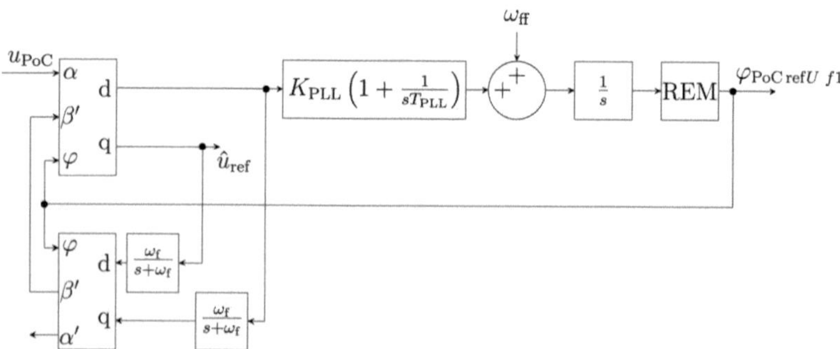

Figure A.3: Modell of an IPT PLL [2].

Most PLLs rely on filtering the input signal, e.g. the voltage at the PoC, and create the orthogonal signal based on the delay that results from the filter function. A simple PLL with an active notch filter (ANF) is the ePLL (see figure A.4). Further synchronization algortihms that are based on filter functions are the second order generalized integrator (SOGI) PLL, the SOGI frequency locked loop (FLL) and the Hilbert PLL.

The Hilbert PLL is based on a finite impulse response (FIR) filter. The coefficients of

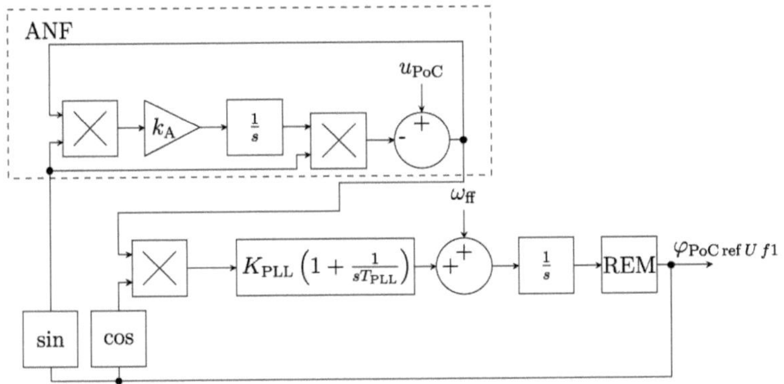

Figure A.4: Modell of an ePLL [2].

that filter can be calculated with

$$h[n] = \begin{cases} 1 - \frac{\cos[(n-0.5N)\pi]}{\pi(n-0.5N)}, & n \neq 0.5N \\ 0, & n = 0.5N \end{cases} \tag{A.2}$$

for each coefficient n and the maximum filter order N, i.e. 31 in this work. The PI controller of the PLL is designed based on the general reference network conditions, i.e. 325 V for $\hat{u}_{\text{PLL}}$ at 50 Hz. The standard design factor 2.4 is chosen, so that the PLL parameters can be calculated in terms of

$$\frac{\omega_{\text{PoC}}(f_1)}{U_{\text{PoC}}(f_1)} = \frac{K_{\text{PLL}}(T_{\text{PLL}}s + 1)}{T_{\text{PLL}}s} \tag{A.3}$$

with

$$T_{\text{PLL}} = 2.4^2 T_{\text{sw}} \tag{A.4}$$

and

$$K_{\text{PLL}} = \frac{1}{2.4\hat{u}_{\text{PLL}} T_{\text{sw}}} \tag{A.5}$$

with the voltage u_{PoC} in the respective frame coordinates and the angular frequency of the voltage at the PoC ω_{PoC}. [16]
The parameters of the PLLs are listed in table A.2.

Table A.2: Parameters of the phase locked loops [2].

PLL type	Parameter	Value
General PLL Properties	Propotional gain K_{PLL}	$10.256\,\frac{\mathrm{Hz}}{\mathrm{V}}$
	Reset time T_{PLL}	$7.2 \cdot 10^{-4}\,\mathrm{s}$
	Feed forward angular frequency ω_{ff}	$2\pi 50\,\mathrm{Hz}$
IPT	ω_{ff}	$2\pi 70.7\,\frac{\mathrm{rad}}{\mathrm{s}}$
ePLL	k_{A}	160

DC-link voltage control

The DC-link voltage control is typically designed as a PI control by the standard formula

$$K_{\mathrm{p\,DC}} = \frac{0.12 C_{\mathrm{DC}}}{T_{\mathrm{sw}}} \tag{A.6}$$

for the proportional gain,

$$T_{\mathrm{n\,DC}} = 17 T_{\mathrm{sw}} \tag{A.7}$$

for the reset time and consequently for the integral gain

$$K_{\mathrm{i\,DC}} = \frac{K_{\mathrm{p\,DC}}}{T_{\mathrm{n\,DC}}}. \tag{A.8}$$

In addition to the DC-link voltage control, a strategy on how to handle the DC-link voltage ripple is often made by the manufacturers. Possible implementations can make use of notch-filtering techniques, either by filtering the signal before its input into the control loop or at the end of the control loop. Another option is to average the reference voltage and feed-forward the original input signal, directly feed-forward the input signal or divide the control output during a normalization by the original input signal. The choice and the individual parameters of the implementation affect the spectrum of the grid-side spectrum. Laboratory measurements and discussions with manufacturers have also indicated that not all manufacturers take countermeasures against the DC-link voltage ripple, namely the 100 Hz component.

AC-current control

The parameters of the AC-current control can be calculated for the PI controller in natural and in dq-frame with

$$K_{\mathrm{p\,AC}} = \frac{L_{\mathrm{f}}}{3\,T_{\mathrm{sw}}}, \tag{A.9}$$

$$K_{\mathrm{i\,AC}} = \frac{R_{\mathrm{f}\,L}}{3\,T_{\mathrm{sw}}} \tag{A.10}$$

and the feed-back gain

$$K_{\mathrm{fb\,AC}} = \frac{K_{\mathrm{p\,AC}}}{K_{\mathrm{i\,AC}}}. \tag{A.11}$$

The control parameters are calculated based on the sum of the inductances and the resistances, i.e. by

$$L_{\mathrm{f}} = L_{\mathrm{f\,g}} + L_{\mathrm{f\,d}} \tag{A.12}$$

and

$$R_{\mathrm{f}\,L} = R_{\mathrm{f\,g}} + R_{\mathrm{f\,d}} \tag{A.13}$$

and depend on the specific filter implementation, e.g. LCL-filter I or LCL-filter II. In case of a simple L-filter, the inductance and resistance values can be directly inserted into (A.9) and (A.10). Also, a PR controller is implemented that is specifically tuned to 50 Hz, 100 Hz and 150 Hz. In addition, a notch filter can be activated to reduce the impact of the interaction of the AC-side filter circuit in its resonance (the LCL-implementations) with the controllers [16]. This notch filter is not required, but works supplementarily to reduce the distortion of the current at the PoC.

Modulator

The modulation algorithm that has been used in this work is the asymmetric unipolar regular sampled PWM with triangular carrier signal [40] as presented in figure A.5 with a zero-order hold (ZOH) block and the bitwise complement (NOT) block. This represents accurately the implemented modulator in real practice that is state of the art in single-phase inverters and converters.

For the sake of academic studies, further algorithms have been implemented according to table A.3.

A rectifier mode has been implemented in addition to the different, implemented modulation algorithms during operation as an inverter for charging the DC-link capacitance when turning on the inverter. If all IGBTs are turned off and the voltage at the PoC is larger than the DC-link voltage, the inverter can operate as a passive rectifier.

A.1.3 Subsystems and masking

The implementation of components can be structured when using the construct of subsystems in MATLAB/Simulink [81]. Each component can be represented by one subsystem. Each subsystem can consist of subsidary subsystems, which enables to implement multiple desgins of the same component. Unused subsystems can be disabled. The variables of disabled subsytems are neglected during the simulation. In addition, local variable handling is possible, i.e. local variables are only visible within the subsystem but not outside the subsystem (i.e. globally). This allows to assign parameters for individual component

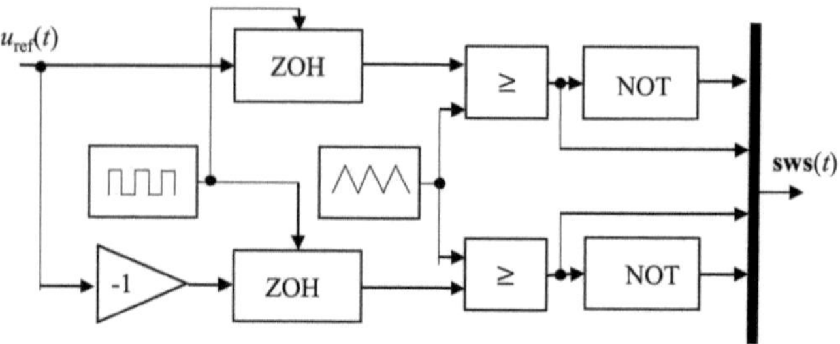

Figure A.5: Scheme of a asymmatric regularly sampled uniplor modulator with triangular carrier signal [5].

topologies with the same name to different component implementations.
The masking function, e.g. the creation of subsystems, facilitates cloning of inverters and slight changes to one inverter desing which is important for large-scale simulations with many inverters.

A.1.4 Component overview

The entire set of implemented inverter components is listed in table A.3 as an overview. A combination of each component implementation together with the other components is possible to operate in simulations thus forming different output characteristics, e.g. currents, at the PoC.

A.1.5 Model configurations

In the presented work, mainly four simulation models with different designs have been studied and compared in [16] to commercially available invertes. The labeling of the models is based on capital Latin letters. Each model can be reassigned to the implemented model components by the alphanumeric code as listed in table A.3.
The corresponding configuration can be read that the first three digits relate to the hardware components each separated by the „x“ and the next four digits relate to the the software components. The dash separates the hardware component numbers from the software component numbers with the numbering from table A.3 and according to the respective scheme by
„AC-side filter-circuit“ x „DC-link capacitance“ x „Switch topology “ - „Grid synchronization“ x „DC-link voltage control“ x „AC-current control“ x „Modulator“.

Table A.3: Component overview of the modular model.

n°	Hardware components			
	AC-side filter circuit	**DC-link capacitance**	**Switch topology**	
1	LCL-filter I	400 µF	Full bridge (H4)	
2	LCL-filter II			
3	L-filter			
	Software components			
	Grid synchronization	**DC-link voltage control**	**AC-current control**	**Modulator**
1	T4D PLL	PI (with notch filter)	PR ($\alpha\beta$)	bipolar natural
2	IPT PLL	PI (no notch filter)	PI (dq)	bipolar regular symmetrical
3	ePLL		PI ($\alpha\beta$)	bipolar regular asymmetrical
4	SOGI PLL			unipolar natural
5	SOGI FLL			unipolar regular symmetrical saw
6	Hilbert PLL			unipolar regular symmetrical triangular
7				unipolar regular asymmetrical
8				rectifier mode

Table A.4: Model configuration from [16].

Label	Configuration
A	1x1x1-1x1x1x7
B	2x1x1-1x1x1x7
C	1x1x1-6x1x1x7
D	1x1x1-1x1x2x7

Linear time-invariant characteristics

The LTI chracteristics (impedance characteristics) of the four simulated inverters are presented in figure A.6. The frequency sweep has been applied with an amplitude of the sweep component of 10 V and a respective phase angle of 0 ° from 100 Hz to 2 kHz in 50 Hz steps for a sinusoidal reference voltage with an amplitude of 325 V at 50 Hz.

A.2 Continuous and discrete component implementation

There are different options to implement a specific inverter component as presented in the following section and published by the autor in [5]. The implementation is based on the model component characteristics and the overall choice of the solver.
To classify model component characteristics two categories can be considered: continuous and discrete characteristics. Often, continuous characteristics relate to physical components, i.e. the hardware. Discrete characteristics are typically related to quantized signals, e.g. due to the clock of a microcontroller, signal processing times in digital environments etc. In practice, simplifications are suitable. However, the implementation and the choice of simplifications affect the choice of solver and their ability and feasibility to enable reasonable simulation results.
It is possible to emulate model components with digital characteristics in continuous model environments in time domain, frequency domain or Laplace domain, if possible delays, e.g. of the sampling, are approximated. [5]
Standard methods for the approximation are the Forward Euler, the Backward Euler and the Tustin method [5]. The Tustin method is mostly used in practice and can be applied by transforming from z-domain into Laplace-domain (s-domain) [82] in terms of

$$s = \frac{2}{T_{\text{sw}}} \frac{z-1}{z+1}. \tag{A.14}$$

To calculate the delay D that results from the overall calculation time of the microcontroller of the control, the sampler and the hold characteristic, the Tustin method according to (A.14) can be applied and leads to

$$D = \frac{e^{-T_{\text{sw}}} - (1 - e^{-T_{\text{sw}}s})}{T_{\text{sw}}}. \tag{A.15}$$

Simplified, D can be approximated by D_{simp} with

$$D_{\text{simp}} = \frac{1}{1.5T_{\text{sw}}s + 1}. \tag{A.16}$$

The consideration of the delay according to the simplification in (A.16) can be applied for the design of the control where the switching time of the IGBTs, i.e. T_{sw} according to (A.1), is much smaller than the time constants of the controlers. The different output signal characteristics and the input signal $u_{\text{ref}}(t)$ based on the different component implementations are visualized in figure A.7.

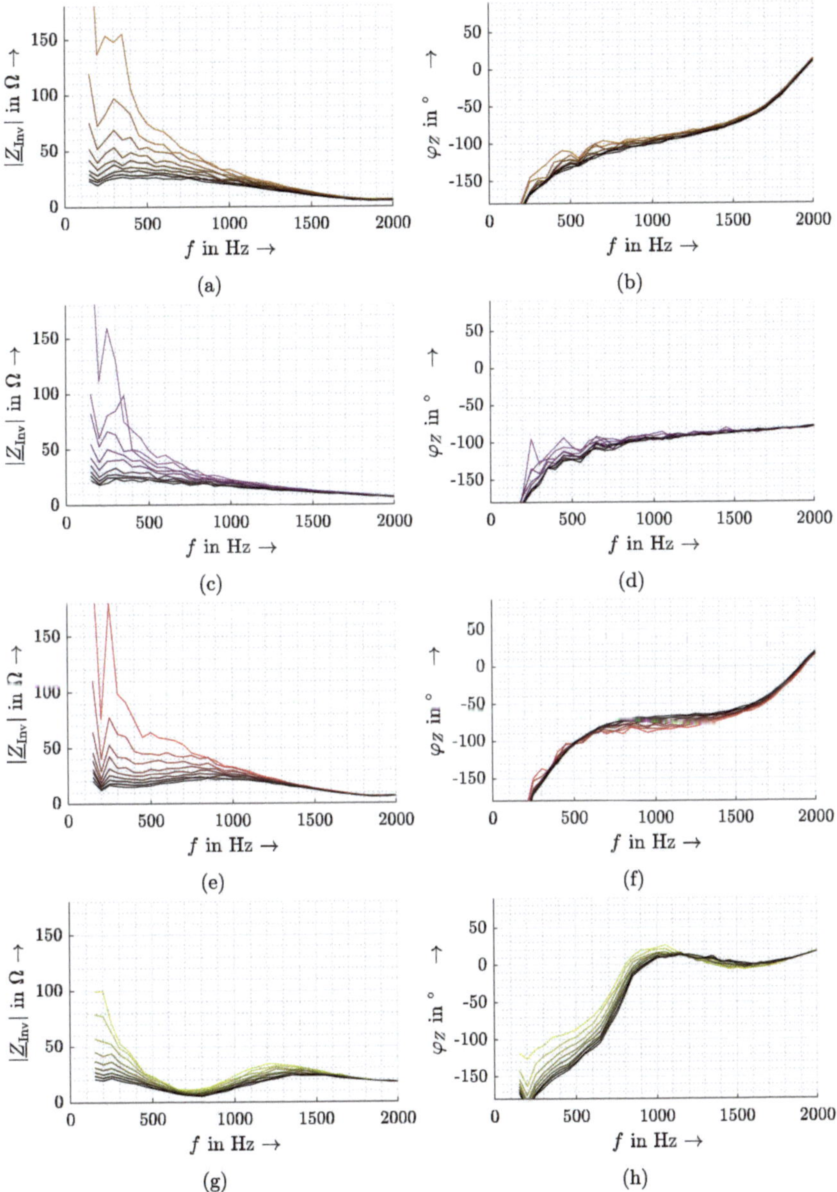

Figure A.6: Magnitude $|\underline{Z}_{\mathrm{Inv}}|$ (left) and phase angle $\varphi_{\mathrm{Inv}\,Z}$ (right) characteristics of inverter A (a, b) inverter B (c, d) inverter C (e, f) and inverter D (g, h) of impedances of inverter models from 500 W (bright color) in 500 W steps to rated power, i.e. 3.9 kW (dark color) based on [16].

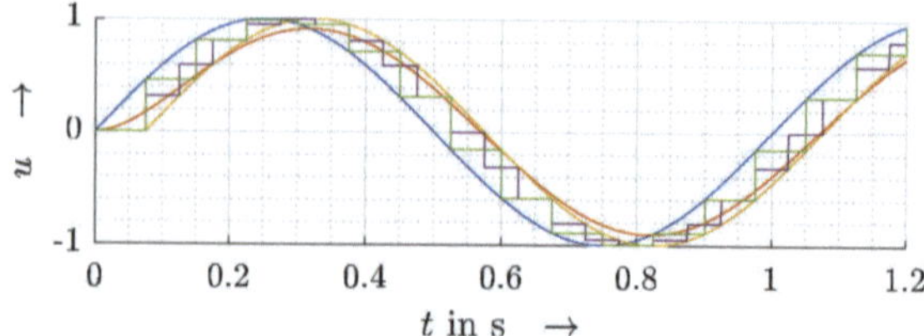

Figure A.7: Time-domain signal of u_{ref} (blue), u_{approx} (red), u_{delay} (yellow), $u_{\mathrm{dig\,SH}}$(green) and $u_{\mathrm{dig\,real}}$ (purple) based on [5].

A.3 Solvers

The following section presents the impact of different solvers for time-domain simulations of an inverter and has been published by the author in [5] by also considering the previous section A.2. The topic is of relevance as it affects not only the simulation duration and the computational power but also the simulation accuracy. Hence, misleading results can cause false conclusions. The software PLECS blockset and MATLAB/Simulink have been compared in [83]. However, it is rather the type of solver that is of interest than the general software, since the solver and respective solver settings define the accuracy and the duration of the simulation results. Most simulation software provides different solvers, though their correct parameter settings are essential for reliable simulation results and left to be set by the user.

A.3.1 Classification

Solvers can be classified into discrete and continuous solvers. Continuous solvers apply numerical integration whereas discrete solvers are more suitable, if discontinuous states are expected during the simulation [5].

A.3.2 Comparison

The following studies different solvers and their settings to validate the overall simulation settings for the simulation-based studies in the presented work.

Simulation environment

For all simulations, a standard office PC system running with Windows 10-64bit has been used. The software runs on an Intel Core i5-6500 @3.20 GHz processor with random-access memory (RAM) of 8 GB. The central processing unit (CPU) score and the memory score are both rated above 8 (Benchmark of Windows Systems). For all simulations, MATLAB/Simulink has been used with the Simscape Power Electrics library. [5]

Simulation procedure

For an exemplary representation of the LV network, two implementations are chosen: an ideal LV network (no impedance) and the reference impedance according to the IEC 60725 [84] as depicted in Figure A.8 based on a network inductance L_g and a network resistance R_g.

At the beginning of the simulation, the DC-link capacitance is charged in rectifier mode. After charging the DC-link capacitance, the power flow is reversed and the PV system injects power into the LV network. The LV network impedance is set to zero (ideal impedance in figure A.8). After 300 ms, the ideal network impedance is switched to the reference network impedance. The reference impedance is connected for 100 ms until switching back to the ideal impedance. [5]

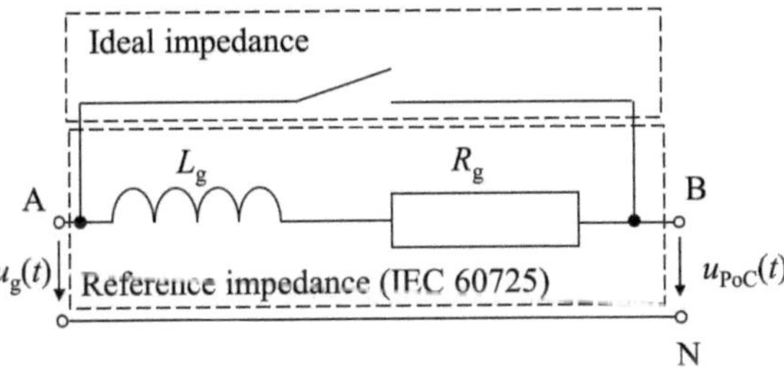

Figure A.8: Scheme of the LV network impedances for the solver study based on [5].

Test cases

As test cases, different solvers with their respective parameters are chosen. No discrete solver has been chosen for the comparison as the simulation based on a discrete solver would also require to change the implementation of certain model components, e.g. the controller implementations would require discrete integrators.

The ode45 solver (test case 1) is a non-stiff solver and used to be the default solver in earlier Simulink versions. The ode23tb solver (test case 2) is a stiff solver. Both solvers are variable step solvers. The relative tolerance is set to 10^{-4} in a first test subcase (a) and to 10^{-7} in a second test subcase (b). For all these test cases, the maximum step size is set to the default solver setting in MATLAB. If the maximum step size does not provide results within the relative tolerance, the step size is reduced and the results are calculated anew for a smaller step size. Therefore, the maximum step size only needs to be large enough thus too large values increase the overall simulation duration.

The ode1 solver is a fixed step solver and parametrized with a fixed step size of $1.25 \cdot 10^{-6}$

Table A.5: Solver parameters [5].

Solver type		Setting		Test case
Variable step solver		**Max. step size**	**Rel. tolerance**	
ode45	Non-stiff	$1.06 \cdot 10^{-6}$	10^{-4}	1a
			10^{-7}	1b
ode23tb	Stiff	$1.06 \cdot 10^{-6}$	10^{-4}	2a
			10^{-7}	2b
Fixed step solver		**Fixed step size**		
ode1		$1.25 \cdot 10^{-6}$		3a
		$1 \cdot 10^{-7}$		3b

and also, for comparison reasons, with a fixed step size of 10^{-7}.
An overview of the solver settings and the test cases is listed in table A.5. [5]

Simulation results

To analyze the results, a DFT has been performed on the current at the PoC for a time window of 20 ms of the simulation part with the connected reference impedance. The results of the DFT are depicted in figure A.9.

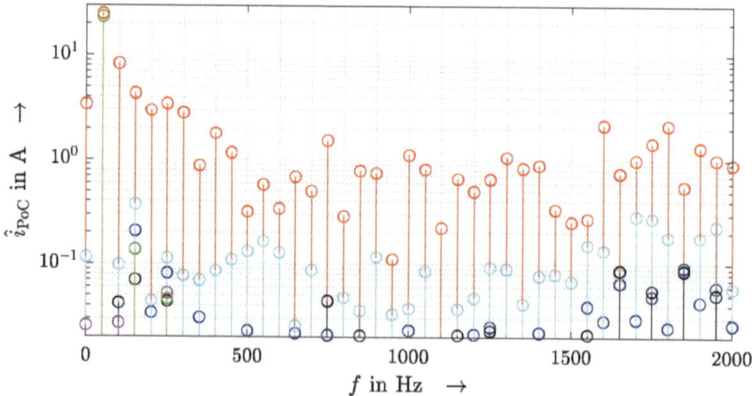

Figure A.9: Current at the PoC with connected reference impedance for test case 1a (red), test case 1b (magenta), test case 2a (black), test case 2b (blue), test case 3a (cyan) and test case 3b (green) based on [5].

In general, a distortion of the current at the PoC is possible and highly likely. However, if the distortion of the current is due to inaccurate solver settings and the current distortion reduces with an increasing solver resolution, a smaller fixed step size can possible improve the accuracy of the simulation results. Thus for testing the solver accuracy, starting with a larger step size and reducing the step size while analyzing the respective signals, a reasonable step size can be identified. For this work, inaccuracies in the range of a few mA are acceptable for the author, e.g. also in relation to the measurement uncertainties in the laboratory. Therefore, for the test cases 1b, 2a, 2b and 3b, the results are considered similar and reasonable [5]. To compare the different test cases, the total distortion current (TDC) is calculated based on all RMS values of the current at the PoC for all harmonic orders including the DC-component, but without the fundamental frequency in terms of

$$\mathrm{TDC} = \sqrt{I_0^2 + \sum_{h=2}^{40} I_h^2}. \tag{A.17}$$

The results for the studied test cases are listed in table A.6. The larger deviations for test case 1a and test case 3a become obvious when comparing the total distortion currents of all simulations.

Table A.6: Total distortion current [5].

Test case	1a	1b	2a	2b	3a	3b
TDC in A	12.6	0.2	0.2	0.2	0.8	0.2

Selected simulation results are visualized in figure A.10 for the voltage of test case 3b and test case 1a. Test case 3b is chosen as reference test case though the visualization of the test cases 1b, 2b and 3a are similar. The largest deviations have been simulated between the results from test case 1a and test case 3b. The ideal switching with infinite slope causes the distinctive voltage drop in figure A.10 (a) and A.10 (b) when switching the reference impedance. A better solver accuracy reduces this voltage drop but leads to a longer simulation duration.

The different distortions for test case 3b and test case 1a become obvious when comparing the detailed voltage plots in figure A.10 (c) and A.10 (d) for the time duration when the reference impedance is connected. The same accounts for a comparison of the currents at the PoC (see figure A.11) and the active and reactive power (see figure A.12).

When only studying the simulation with the former default solver parameters, i.e. the ode45 solver, the simulation results would possibly indicate an instable operation of the inverter. Even the active and reactive power would indicate such an inverter behavior besides the obvious current and voltage distortion. As demonstrated, increasing the relative tolerance points out the misleading results. Another proof for the inaccurate simulation results of test case 1a can be achieved by the formal analysis that indicates a stable operation of the inverter.

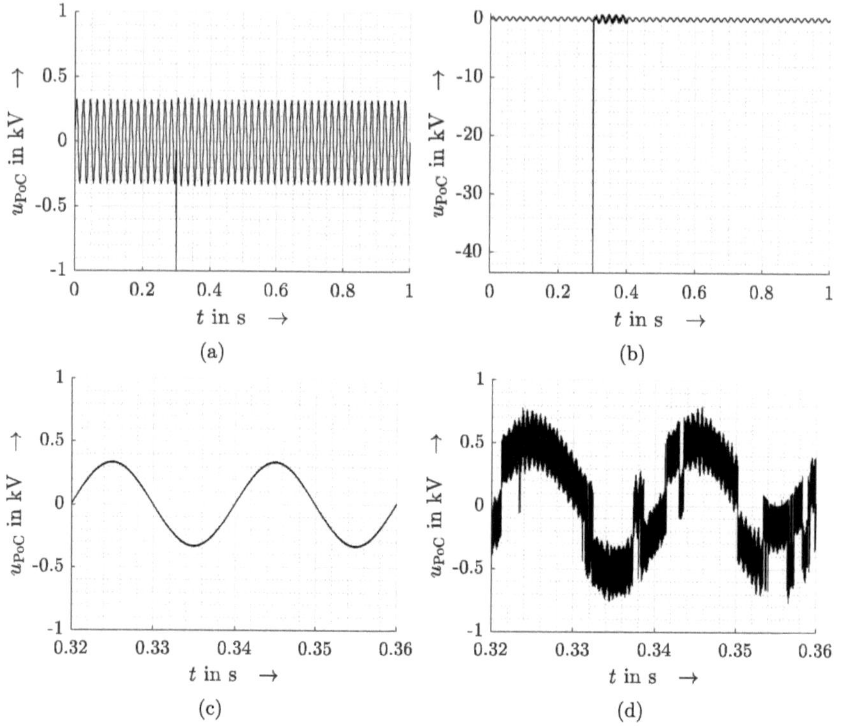

Figure A.10: Voltage u_{PoC} and the relevant detail for test case 3b (a, c) and test case 1a (b, d) based on [5].

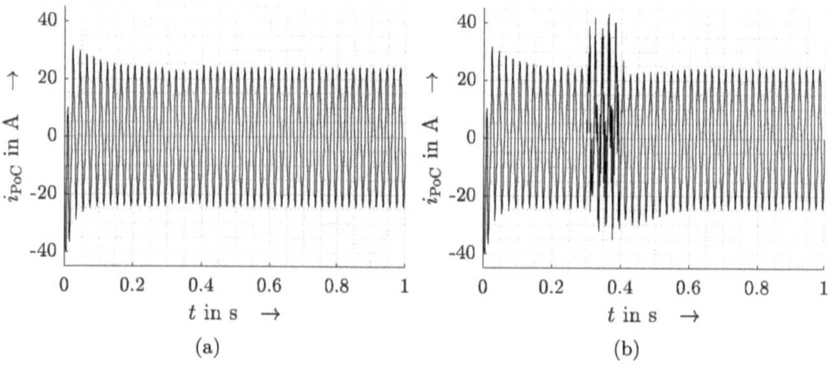

Figure A.11: Current i_{PoC} for test case 3b (a) and 1a (b) based on [5].

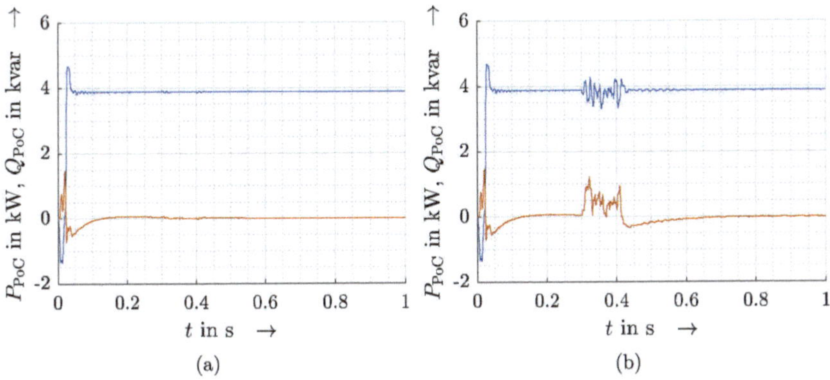

Figure A.12: Active power P_{PoC}(blue) and reactive power Q_{PoC}(red) for test case 3b (a) and test case 1a (b) based on [5].

Simulation time

Power electronic systems are typically based on different time constants, e.g. with regard to the controllers and the grid synchronization. Since this reflects a stiff problem, stiff solvers perform typically better than non-stiff solvers. Consequently, the stiff solver in test case 2 performs better than the non-stiff solver in test case 1 (see also table A.7). The fixed step solver simulates the switching process faster than the variable step solvers since it only consideres the fixed step size while the variable step solvers adapt the step size to calculate a solution in detail for the switching process. For the fixed step solver, the appropriate fixed step size needs to be identified based on multiple simulations, e.g. as in test case 3a and 3b, to ensure a suitable accuracy. The simulation durations for all test cases are listed in table A.7. [5]

Table A.7: Simulation duration [5].

Test case	1a	1b	2a	2b	3a	3b
Simulation duration	3885 s	5151 s	287 s	474 s	46 s	664 s

A.3.3 Conclusions on choice of solver

The solvers and the respective parameters affect the accuracy and the duration of the simulation. Stiff solvers are typically the better choice though initial tests should be performed to ensure appropriate solver settings. In general, the choice of solvers depends on the specific problem so that for the studied objective, fixed time step solvers can possbily present a suitable option. Furthermore, already during the implementation process, it should be considered that stiff and non-stiff solvers can be exchanged but discrete and continuous solvers require different implementations thus a later switch from discrete to continuous solvers can require a re-implementation of model parts. Inappropriate solver settings can lead to false conclusions, especially regarding the voltage and current distortion at the PoC and the stable or rather instable operation of the inverter.

Appendix B

Test stand

The test stand is presented with regard to the devices that form the test stand and the setup of the test stand environment and the measurement devices that are used to record the currents and voltages.

B.1 Test stand devices

B.1.1 Programmable power amplifier

The programmable power amplifier can emulate the LV network and is rated 45 kVA. It can be used as a voltage source, e.g. to represent the background voltage, by providing either a single- or a three-phase voltage.

Input impedance characteristic

The input impedance characteristics of the grid simulator for the frequency-dependent magnitude and the phase angle are depicted in figure B.1.

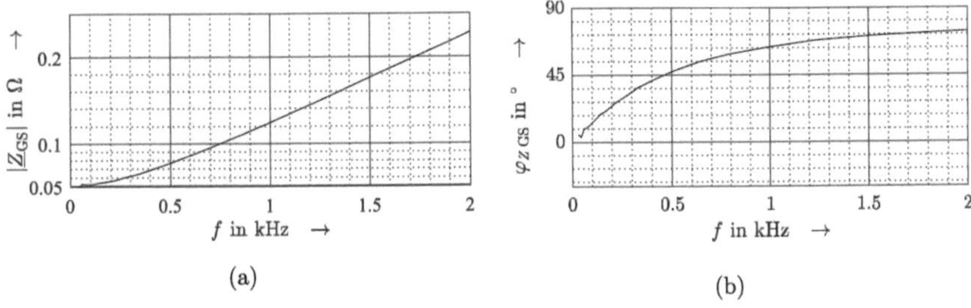

Figure B.1: Input impedance characteristics of the grid simulator for magnitude (a) and phase angle (b) based on [12].

B.1.2 DC power source

The DC power source provides a DC power up to 7 kW or rather 10 kW [85]. Based on the laboratory setup, it has been typically used with a rated power of 7 kW as this configuration has been sufficient for all measurements.
Via the respective software, it is possible to set a power trajectory to emulate typical PV string characteristics, i.e. the voltage-current characteristics of the PV panels (also called PV modules). In theory, the number of parallel strings and the number of PV panels per string can be set. For this work, a standard setup with one string and one type of PV panels has been used for all measurements. Technically, the emulation of different PV modules connected to up to six PV strings is possible, though this is only important for PV power plants of larger size.

B.1.3 Surge protection devices

For the surge protectors, devices of type 2049-60-BLF Bourns [86] with a voltage rating of 600 V for peak impulse currents up 15 kA are implemented.

B.2 Measurement devices

In the following, the measurement devices are introduced. The overall measurement uncertainties are disclosed in section B.2.4 though device-specific measurement uncertainties are not further dicussed.

B.2.1 Current clamps

Current measurements have been performed with current clamps of type PNA-Clamp-150-DC [87]. Rogowsky coils have been available, but due to a possible presence of DC currents, the use of rogowsky coils is not recommended by the author.

B.2.2 Voltage sensors

For an A/D transform and for direct voltage measurements, HSI-LV modules [88] and HSI-high voltage (HV) modules [89] are used in the transient recorder for all measurements.

B.2.3 Transient recorder

The transient recorder Dewetron Dewe-2600 [90] is used for all measurements. It provides a sampling rate of up to 10 Msamp/s for each channel.

B.2.4 Measurement uncertainties

The measurement uncertainty assessment is applied to the entire measurement chain. The identified measurement uncertainty is not only covering the impact of the measurement devices, but also of the recording, e.g. in terms of the sampling rate (1 MHz), and a butterworth filter of 3rd order at a cutoff frequency of 300 kHz that is implemented in the digital signal processing chain of the transient recorder. The voltage and current uncertainty is evaluated based on the deviations that are calculated in relation to an injected current and an applied voltage. The device that injects the currents and applies the voltages is a calibrated CMC256plus [91] that has an output accuracy of better than 1 % for current values above 62.5 µA and voltage values above 7.5 mV. Based on the high output accuracy of the CMC256, the error made by the CMC256 is neglected.
The uncertainty of the current measurements are calculated for the measurement range of 25 A in terms of

$$\Delta\hat{i}_{25\,\mathrm{A}}(\hat{i}_{\mathrm{ref}}) = |\hat{i}_{25\,\mathrm{A}} - \hat{i}_{\mathrm{ref}}| \tag{B.1}$$

$$\Delta\hat{i}_{\mathrm{rel}\,25\,\mathrm{A}}(\hat{i}_{\mathrm{ref}}) = \frac{|\hat{i}_{25\,\mathrm{A}} - \hat{i}_{\mathrm{ref}}|}{\hat{i}_{\mathrm{ref}}} \tag{B.2}$$

and

$$\Delta\varphi_{25\,\mathrm{A}}(\hat{i}_{\mathrm{ref}}) = |\varphi_{25\,\mathrm{A}}(\hat{i}_{\mathrm{ref}}) - \varphi_{i\,\mathrm{ref}}(\hat{i}_{\mathrm{ref}})| \tag{B.3}$$

and for the measurement range of 50 A respectively. The results are depicted in figure B.2.
Similarly, the uncertainties of the voltage measurements can be calculated in terms of

$$\Delta\hat{u}_{400\,\mathrm{V}}(\hat{u}_{\mathrm{ref}}) = |\hat{u}_{400\,\mathrm{V}} - \hat{u}_{\mathrm{ref}}| \tag{B.4}$$

$$\Delta\hat{u}_{\mathrm{rel}\,400\,\mathrm{V}}(\hat{u}_{\mathrm{ref}}) = \frac{|\hat{u}_{400\,\mathrm{V}} - \hat{u}_{\mathrm{ref}}|}{\hat{u}_{\mathrm{ref}}} \tag{B.5}$$

$$\Delta\varphi_{400\,\mathrm{V}}(\hat{u}_{\mathrm{ref}}) = |\varphi_{400\,\mathrm{V}}(\hat{u}_{\mathrm{ref}}) - \varphi_{u\,\mathrm{ref}}(\hat{u}_{\mathrm{ref}})| \tag{B.6}$$

and for 800 V respectively. The results are depicted in figure B.3.

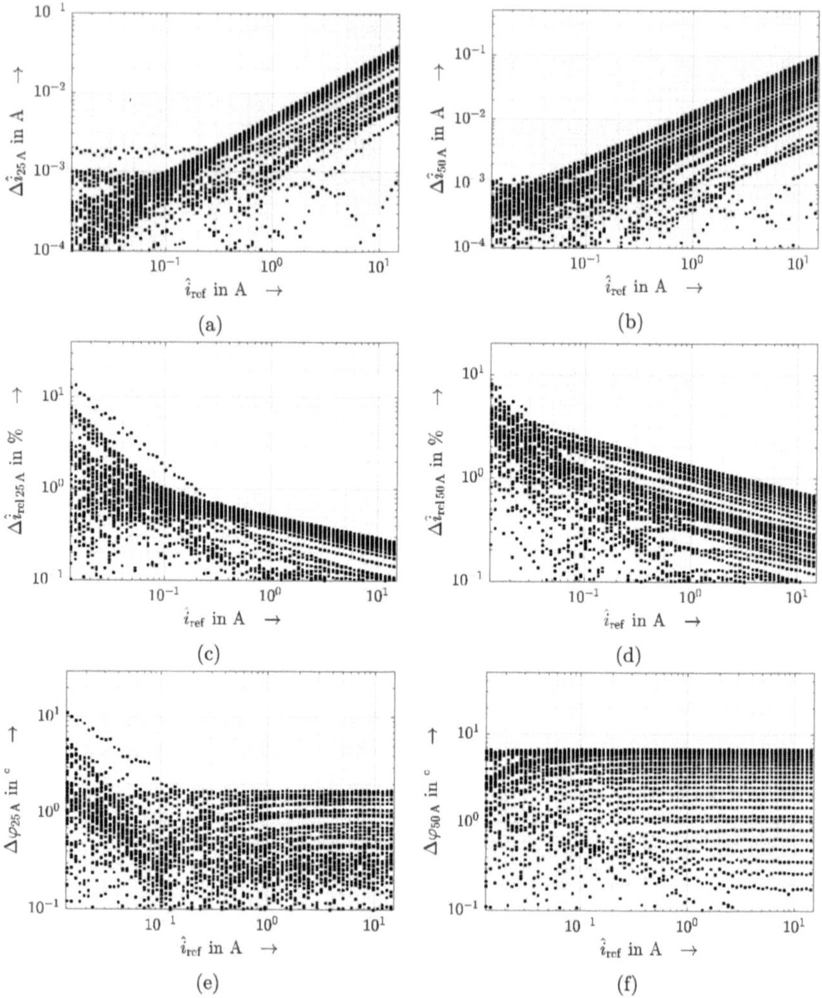

Figure B.2: Measurement uncertainty for amplitude and phase angle for the 25 A (a, c, e) and the 50 A (b, d, f) measurement range.

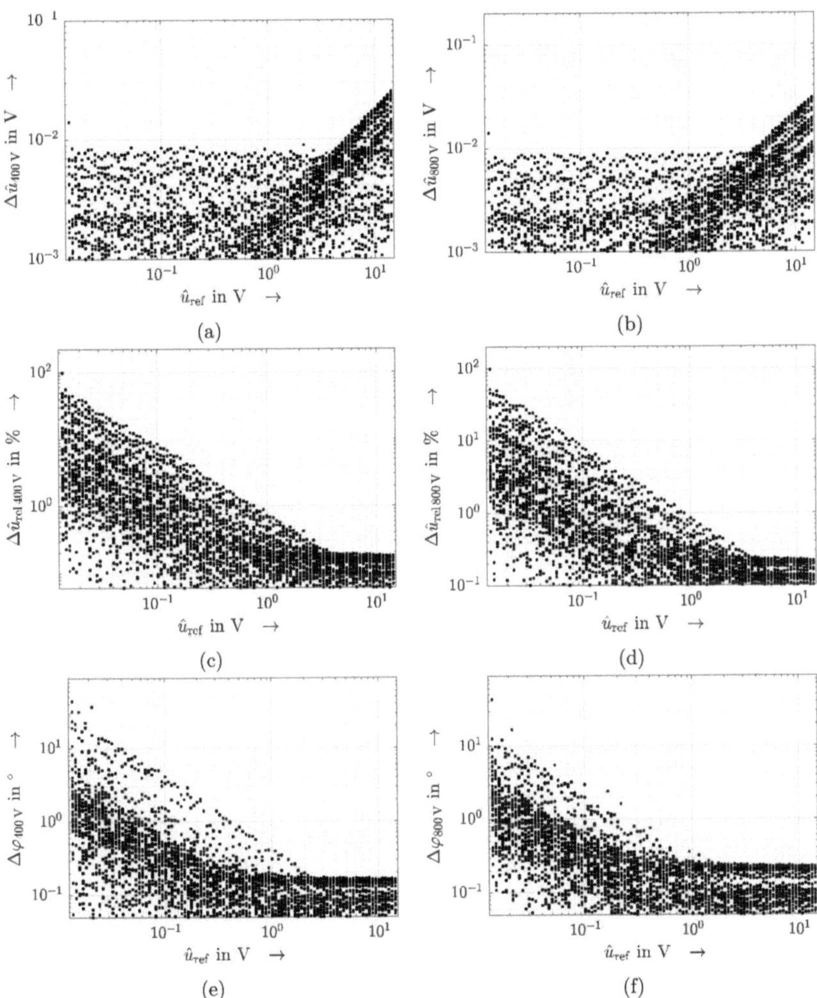

Figure B.3: Measurement uncertainty for amplitude and phase angle for the 400 V (a, c, e) and the 800 V (b, d, f) measurement range.

Appendix C

Low voltage network characteristics

C.1 Background distortion

The frequency spectrum of a pointed-top voltage waveform and a flat-top voltage waveform are listed in table C.1 and table C.2 and depicted in figure C.1 according to [53]. The flat-top voltage waveform represents a characteristic voltage waveform in public LV networks and the pointed-top voltage waveform represents a typical voltage waveform in industrial LV networks.

Table C.1: Spectrum of a flat-top voltage waveform [53].

Frequency in Hz	Amplitude in V	Phase angle φ_U in °
50	229.8924	0
150	5.4501	0
250	3.8271	180
350	2.0398	0
450	0.5651	180
550	0.3084	180
650	0.5567	0
750	0.3751	180
850	0.0502	0
950	0.1820	0
1050	0.2225	180
1150	0.1115	0
1250	0.0373	0
1350	0.1214	180
1450	0.1053	0
1550	0.0243	180

Table C.2: Spectrum of a pointed-top voltage waveform [53].

Frequency in Hz	Amplitude in V	Phase angle φ_U in °
50	229.895	0
250	6.621	0
350	4.73	180
450	1.44	180

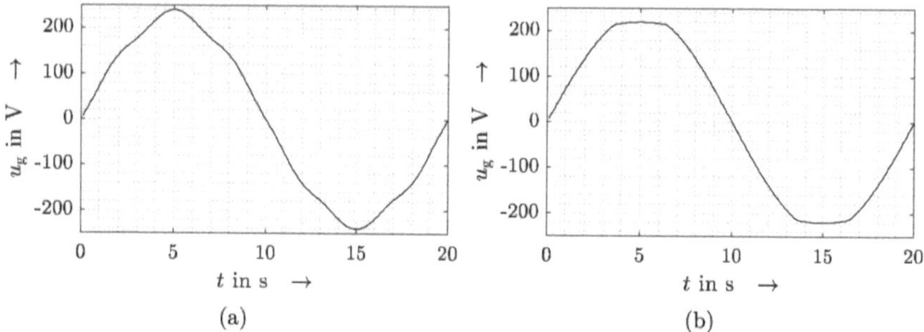

Figure C.1: Waveform of a pointed-top (a) and a flat-top (b) voltage profile according to [53].

C.2 Network impedance characteristics

The network impedances have been measured in rural and urban public LV networks in central Europe, i.e. Germany, Switzerland, Austria and the Czech Republic [70]. The measurement method is based on a frequency sweep. Compared to the frequency sweep performed to identify grid-connected devices, the network impedance is identified by injecting a current into the grid instead of applying a voltage at the clamps. The current source, a linear amplifier, performs the sweep while the voltage response is measured. The injected current is limited, so that the resulting frequency-specific change of the voltage is below 1 V. This is ensured by a pre-measurement that is applied to define the exact current amplitude for the sweep.

It is assumed that the network configuration remains the same (no significant switching of components within the network) during the measurement.

For the presented network impedances, the sweep is performed up to 39 kHz.

About 80 % of the network impedances have a resonance peak between 600 Hz and 1.5 kHz and the overall maximum resonance rise factor is 1.4 [1].

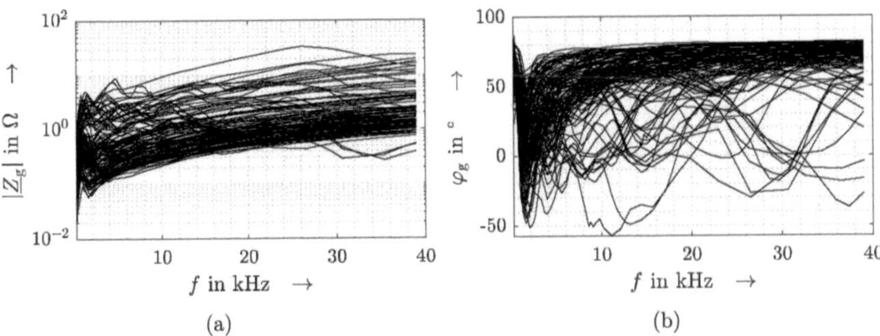

Figure C.2: Magnitude (a) and phase angle (b) characteristics of public LV network impedances measured in Germany, Austria, Switzerland and the Czech Republic based on [70].

Appendix D

Component identification

The most elegant way to describe commercially available devices is by developing grey-box models. In a best case scenario, all information about an individual device can be included in a suitable model structure. The model and the respective knowledge can possibly be used for simulations, design analysis and monitoring. Before this model structure can be developed in general for a device, it is necessary to provide approaches that identify information about the inverter components to upgrade the model from the black-box model to a grey-box model.
The identification and an equivalent representation of model components is therefore of interest. Hence, the author has developed an approach to identify the AC-side filter circuit as published in [8] and an approach to identify the DC-link capacitance as published in [9] that are both described in the following of this section accordingly. The requirement for the applicability on commercially available devices is that the measurements have to be performed non-invasively. Non-invasiveness implies that though point measurements, i.e. potential measurements on the circuit board, can be performed inside the inverter, all currents are only measured at external clamps, i.e. the AC-side terminals or the DC-side terminals.

D.1 AC-side filter circuit

To identify the AC-side filter circuit, the two-port theory is applied as explained in the following based on [8].

D.1.1 Theory

Two-port network analysis

Two-port networks represent a special case of four-pole networks. If the port condition is fulfilled, the four-pole qualifies for the port network theory. The port condition requires that each two terminals, i.e. poles of a network, can be represented as a port if the current into one terminal is equal to the negative current of the other terminal [92]. For two-port networks, four parameters have been introduced, namely the impedance, the admittance, the hybrid and the inverse hybrid parameters and respective methods to identify these parameters are based on established state of the art approaches. For the harmonic analysis, typically the current of the DUT at the PoC is of interest, e.g. compared to circulating currents within the filter. Under these circumstances and with regard to the general description of the admittance parameters $\underline{Y}_{11}$, $\underline{Y}_{12}$, $\underline{Y}_{21}$ and $\underline{Y}_{22}$ by

$$\begin{pmatrix} \underline{I}_1(f) \\ \underline{I}_2(f) \end{pmatrix} = \begin{pmatrix} \underline{Y}_{11}(f) & \underline{Y}_{12}(f) \\ \underline{Y}_{21}(f) & \underline{Y}_{22}(f) \end{pmatrix} \begin{pmatrix} \underline{U}_1(f) \\ \underline{U}_2(f) \end{pmatrix} \tag{D.1}$$

only the identification of $\underline{Y}_{11}$ and $\underline{Y}_{12}$ is of relevance. The voltages and currents are defined according to figure A.1. It is not necessary and in practice also often not possible to measure the current $\underline{I}_2$ on the circuit board. It is still required to measure the voltages $\underline{U}_1$ and $\underline{U}_2$ with

$$\underline{Y}_{11}(f) = \left.\frac{\underline{I}_1(f)}{\underline{U}_1(f)}\right|_{\underline{U}_2(f)=0} \tag{D.2}$$

and

$$\underline{Y}_{12}(f) = \left.\frac{\underline{I}_1(f)}{\underline{U}_2(f)}\right|_{\underline{U}_1(f)=0}. \tag{D.3}$$

According to classical identification approaches, each side of the two-port network has to be short-circuited one by one. For commercial devices, the filter circuit cannot be disconnected and furthermore, for grid-following inverters, a voltage at the AC-side terminals has to be provided to operate the device. Thus, $\underline{U}_1$ cannot be short-circuited as required in (D.3) and $\underline{U}_2$ cannot be applied directly and also the respective clamps (where $\underline{U}_2$ is applied) cannot be short-circuited. The identification method has to be adapted and standard *RLC*-meters cannot be used but,as already mentioned, potential measurements on the circuit board are possible. [8]

Identification method

The newly proposed identification method makes use of the knowledge of the generation of the frequency coupling components, i.e. the analysis in chapter 3. If the device is operating and a frequency sweep is performed at the AC-side terminals, the inverter generates frequency coupling components in the current response. The filter circuit is assumed virtually linear in the studied frequency range, i.e. up to 2 kHz. In theory, all other components of the inverter, e.g. control algorithms, could be nonlinear.
The identification approach is performed in two steps: firstly, the identification of the frequency coupling elements and secondly, the calculation of $\underline{Y}_{11}$ and $\underline{Y}_{12}$. For two-pole representations, e.g. the black-box representation of the inverter seen from the grid-

side, only the AC-side voltage $\underline{U}_1$ is present and the device-side voltage $\underline{U}_2$ is an internal parameter. By opening the inverter case, frequency components in $\underline{I}_1$ that are not present in $\underline{U}_1$ can be measured in $\underline{U}_2$. For all AC-side current frequency components for which the AC-side voltage is zero, the admittance $\underline{Y}_{12}$ can be calculated according to (D.3) hence all frequency coupling currents at non-excited voltage frequencies can be used. Averaging can be applied on all values of $\underline{Y}_{12}$ that have been calculated more than once. With the identification of $\underline{Y}_{12}$ and the measurements of $\underline{U}_1$, $\underline{U}_2$ and $\underline{I}_1$, it is possible to calculate $\underline{Y}_{11}$ in terms of

$$\underline{Y}_{11}(f) = \frac{(\underline{I}_1(f) - \underline{U}_2(f)\underline{Y}_{12}(f))}{\underline{U}_1(f)}\Big|_{\underline{U}_1 \neq 0}. \tag{D.4}$$

The frequency-dependent characteristics can be smoothed by applying a smoothing function, specifically if the admittance characteristic is assumed to show no sudden changes in its frequency characteristics (magnitudes and phase angles). The smoothing function can be described as a moving average filter considering five values at respective frequencies at harmonic order h by

$$\underline{Y}_{\mathrm{smooth}}(f_h) = \frac{\underline{Y}(f_{h-2}) + \underline{Y}(f_{h-1}) + \underline{Y}(f_h) + \underline{Y}(f_{h+1}) + \underline{Y}(f_{h+2})}{5} \tag{D.5}$$

to reduce the effect of the measurement uncertainties. [8]

D.1.2 Simulative validation

Reference calculation

To validate the previously described theory, a time-domain simulation model with known parameters and the LCL-filter I (see section A.1.1) is used. As reference for the validation of the identification method, the transfer admittances $\underline{Y}_{11}$ and $\underline{Y}_{12}$ are calculated based on the exact implemented filter circuit elements with the device-side filter impedance $\underline{Z}_{\mathrm{f\,d}}$ by

$$\underline{Z}_{\mathrm{f\,d}}(f) = j2\pi f L_{\mathrm{f\,d}} + R_{\mathrm{f\,d}} \tag{D.6}$$

the grid-side filter impedance $\underline{Z}_{\mathrm{f\,g}}$ by

$$\underline{Z}_{\mathrm{f\,g}}(f) = j2\pi f L_{\mathrm{f\,g}} + R_{\mathrm{f\,g}}, \tag{D.7}$$

and the T-branch filter impedance $\underline{Z}_{\mathrm{f}\,C}$ by

$$\underline{Z}_{\mathrm{f\,C}}(f) = \frac{-j}{2\pi f C_{\mathrm{f}}} + R_{\mathrm{f}\,C}. \tag{D.8}$$

The transfer admittances result in

$$\underline{Y}_{11\,\mathrm{an}}(f) = \frac{1}{\underline{Z}_{\mathrm{f\,g}}(f) + \frac{\underline{Z}_{\mathrm{f\,d}}(f)\underline{Z}_{C\,\mathrm{f}}(f)}{\underline{Z}_{\mathrm{f\,d}}(f)+\underline{Z}_{C\,\mathrm{f}}(f)}} \tag{D.9}$$

and

$$\underline{Y}_{12\,\mathrm{an}}(f) = \frac{1}{\underline{Z}_{\mathrm{f\,d}}(f) + \frac{\underline{Z}_{C\,\mathrm{f}}(f)\underline{Z}_{\mathrm{f\,g}}(f)}{\underline{Z}_{C\,\mathrm{f}}(f)+\underline{Z}_{\mathrm{f\,g}}}} \cdot \frac{\frac{\underline{Z}_{C\,\mathrm{f}}(f)\underline{Z}_{\mathrm{f\,g}}}{\underline{Z}_{C\,\mathrm{f}}(f)+\underline{Z}_{\mathrm{f\,g}}}}{\underline{Z}_{\mathrm{f\,g}}}. \tag{D.10}$$

The analytically calculated frequency domain characteristics (index „an“) of $\underline{Y}_{11\,\mathrm{an}}$ (blue solid lines) and $\underline{Y}_{12\,\mathrm{an}}$ (red solid lines) are depicted in figure D.1 based on [8].

Simulative method application

The values obtained by the identification based on the simulative (index „sim“) measurements are presented by stars (∗) and plus (+) signs. According to chapter 3, each frequency component of $\underline{U}_{\mathrm{PoC}}$ at any frequency f_n causes a multi-frequent current response at $f_n \pm 2mf_1$. This current response decreases for an increasing m. Consequently, only the current response at frequencies at $f_n - 2f_1$ by means of $\underline{Y}_{12\,\mathrm{sim}-}$ and at $f_n + 2f_1$ by means of $\underline{Y}_{12\,\mathrm{sim}+}$ are evaluated. By averaging over $\underline{Y}_{12\,\mathrm{sim}-}$ and $\underline{Y}_{12\,\mathrm{sim}+}$ at values for the same frequencies, $\underline{Y}_{12\,\mathrm{sim}}$ can be calculated.
Using both values for averaging and starting at 100 Hz for the evaluation of the first value at the $-2f_1$-component leads to 300 Hz for the $+2f_1$-component and consequently also to 300 Hz as first value for $\underline{Y}_{12\,\mathrm{sim}}$. Similarly, the maximum values of the admittance with respect to the studied frequency range up to 2 kHz are at 1800 Hz. More generally related to the measured frequency range, the identifiable admittance range is narrowed down by $4f_1$ at the upper and lower frequency limits. Though for this study, only the harmonic frequency range is of interest, the studied frequency range can be extended to higher frequencies.
In figure D.1, all filter admittances, i.e. the analytically calculated, the simulation-based and also the two-pole (TP) representation, are depicted.
At can be concluded that the simulation results are resonable, comparing them with the analytically calculated values.
It is possible to reduce the steps of the applied frequency sweep (50 Hz steps) to achieve a better resolution and consequently improve the results of the smoothing.
The relative error $\mathrm{err}_{12\,\mathrm{sim}}$ can be calculated for the averaged admittance $\underline{Y}_{12\,\mathrm{sim}}$ with

$$\mathrm{err}_{12\,\mathrm{sim}}(f) = \frac{|\underline{Y}_{12\,\mathrm{sim}}(f)| - |\underline{Y}_{12\,\mathrm{an}}(f)|}{|\underline{Y}_{12\,\mathrm{an}}(f)|}. \tag{D.11}$$

The results are shown in figure D.2. The accuracy of the results is depending on the solver accuracy (see appendix A.3). The achieved accuracy of the simulation-based method is

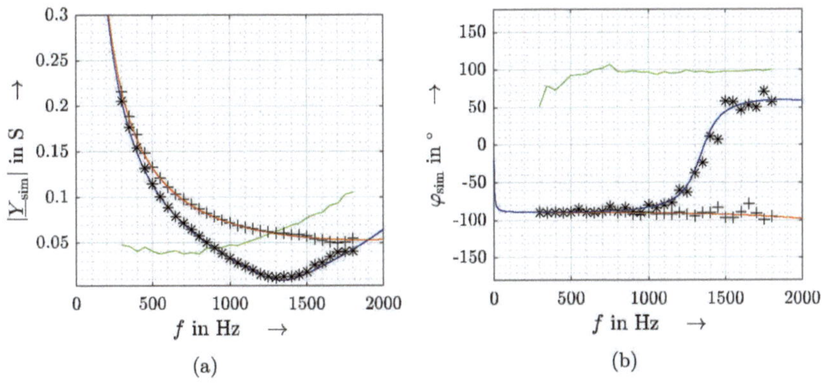

Figure D.1: Magnitude (a) and phase angle (b) characteristics of $\underline{Y}_{11\,\mathrm{an}}$ (blue line), $\underline{Y}_{11\,\mathrm{sim}}$ (black *), $\underline{Y}_{12\,\mathrm{an}}$ (red line), $\underline{Y}_{12\,\mathrm{sim}}$ (black +) and $\underline{Y}_{\mathrm{TP\,sim}}$ (green line) in the frequency range up to 2 kHz based on [8].

in the range of 10^{-3} S, which is better than the measurement accuracy (see appendix B.2.4). The relative error depicted in figure D.2 demonstrates the challenges to identify admittances with small magnitudes, e.g. $\mathrm{err}_{12\,\mathrm{sim}+}$ at 1750 Hz with the maximum relative error of 35.5 %, and more general an increasing relative error with an increasing frequency and a decreasing admittance magnitude value. [8]

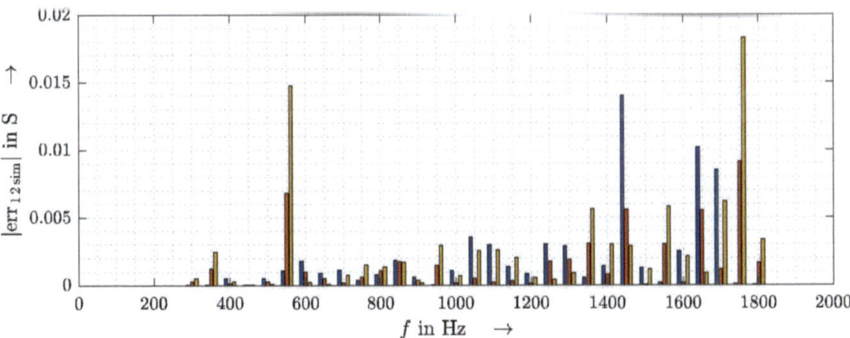

Figure D.2: Relative error $\mathrm{err}_{12\,\mathrm{sim}+}$(yellow), $\mathrm{err}_{12\,\mathrm{sim}-}$ (blue) and $\mathrm{err}_{12\,\mathrm{sim}-}$(red) of the simulation results based on [8].

D.1.3 Measurements

The previous simulation has validated the theory. In the following, measurements are performed and presented for a commercially available inverter (index „meas“). The detailed design parameters including the filter circuit parameters are not known. Therefore, the achieved results cannot be compared to theoretic values for a purpose of validation.

As annotated previously, the measurements require the operation of the inverter with an open case as shown for the studied inverter in figure D.3. As described previously, the voltages $u_1(t)$ and $u_2(\mathrm{t})$ have been measured as well as the current $i_1(t)$ through the clamps with voltage $u_1(\mathrm{t})$. The achieved measurement results are based on the frequency

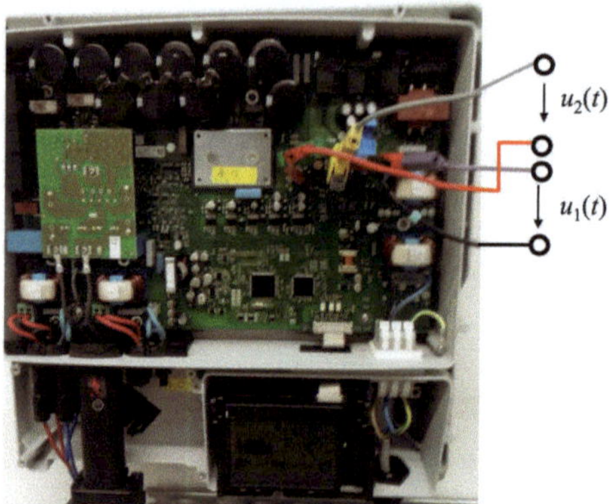

Figure D.3: Commercial PV inverter with open case [8].

sweep with a sinusoidal reference voltage with an amplitude of 325 V at 50 Hz and distortion components with 10 V amplitude and 0° as sweep parameters. For the frequency sweep, a step size of the distortion component of 10 Hz is chosen for a better resolution in frequency domain though this comes at the cost of a longer measurement duration.

The averaged admittance characteristics for $\underline{Y}_{12\,\mathrm{meas}}$ calculated based on $\underline{Y}_{12\,\mathrm{meas+}}$ and $\underline{Y}_{12\,\mathrm{meas-}}$ and the linearized two-port equivalent of the inverter $\underline{Y}_{\mathrm{TP\,meas}}$ are shown in figure D.4.

The values with the index „smooth“ are the values that are calculated based on the smoothing function in (D.5).

D.1.4 Advantages

The main purpose of the AC-side filter circuit identification method is to make grey-box modeling possible in the future. In addition, some further aspects are outlined in the next subsections.

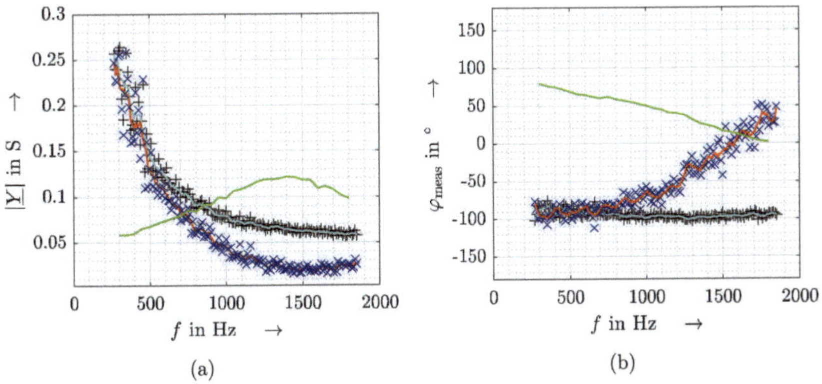

Figure D.4: Magnitude characteristics (a) and phase angle characteristics (b) of $\underline{Y}_{11\,\mathrm{meas}}$ (blue x), $\underline{Y}_{11\,\mathrm{meas\,smooth}}$ (red line), $\underline{Y}_{12\,\mathrm{meas}}$ (black +), $\underline{Y}_{12\,\mathrm{meas\,smooth}}$ (cyan line) and $\underline{Y}_{\mathrm{TP\,meas}}$ (green line) in the frequency range up to 2 kHz based on [8].

Consideration of manufacturing tolerances

The lack of reference values makes a final comparison impossible. However, in practice, circuit elements can differ largely from the specified values in data sheets due to manufacturing tolerances. Typical manufacturing tolerances ε, e.g. according to [93], are listed in table D.1 with the parameter definitions according to figure A.1.

Table D.1: Exemplary manufacturing tolerances based on [8].

Parameter	Value
ε_{RfC}	0.15
ε_{Cf}	0.2
ε_{Lfg}	0.15
ε_{Rfg}	0.5
ε_{Lfd}	0.15
ε_{Rfd}	0.5

Furthermore, the frequency dependency of individual circuit elements is not disclosed. It can be convenient to reduce the uncertainties of the data sheet parameters by applying the proposed identification method. The method also includes parasitic inductances, resistances and capacitances that can result from connectors and also magnetic couplings. Based on the originally intended circuit element parameters, the manufacturer can ensure the production quality. In addition, the method can be applied for device studies, e.g. to identify a failure of circuit elements and aging, without the need to remove the filter circuit from the rest of the device and is applied during operation.

For the manufacturing tolerances listed in table D.1, a 3σ-range (i.e. 98 % environment) around the parameter values is assumed and can be calculated with

$$R_{C\,3\sigma} = R_C(1 \pm \varepsilon_{RC}) \tag{D.12}$$

and respectively for all other circuit element parameters. This normal distribution can be considered in a Monte Carlo (MC) simulation with the simulation model from subsection D.1.2 and the tolerances from table D.1. The results for the admittance magnitudes of the MC are shown in figure D.5 (a) for 5000 iterations. In figure D.5 (b), the relative deviations $\Delta|\underline{Y}_{\mathrm{rel}\,11}|$ of $|\underline{Y}_{11\,\mathrm{an}}|$ and respectively $\Delta|\underline{Y}_{12\,\mathrm{an}}|$ of $|\underline{Y}_{12\,\mathrm{an}}|$ are presented. The relative deviation for $|\underline{Y}_{11}|$ becomes maximal at the resonance minimum [8].
The manufacturing tolerances have an impact on the amplification at the resonance and on the resonance frequency. Simply using the data sheet parameters can cause large variations compared to the real values, e.g. identified by measurements.

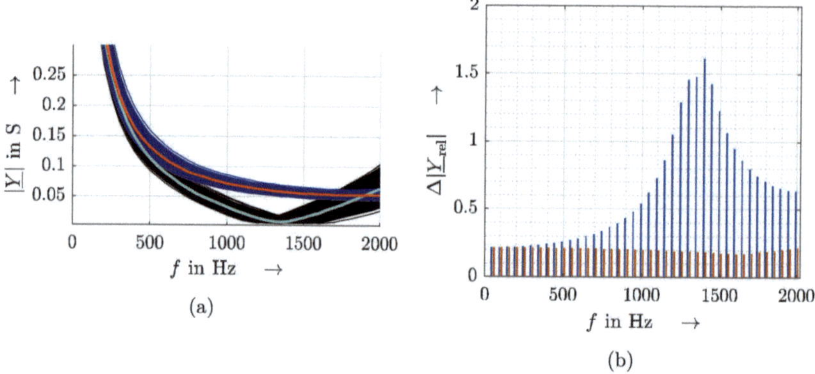

Figure D.5: Magnitude values $|\underline{Y}_{11\,\mathrm{an\,MC}}|$ (black), $|\underline{Y}_{11\,\mathrm{an}}|$ (cyan), $|\underline{Y}_{12\,\mathrm{an}}|$ (red) and $|\underline{Y}_{12\,\mathrm{an\,MC}}|$ (blue) in (a) and relative deviations $\Delta|\underline{Y}_{\mathrm{rel}\,11}|$ (blue) and $\Delta\underline{Y}_{\mathrm{rel}\,12}|$ (red) in (b) from the Monte Carlo simulation with 5000 iterations [8].

Advances of identification of harmonic contribution

The representation of the filter circuit as a four-pole (i.e. the two-port model) enables the identification of the harmonic contribution of the device on a grid-connected device. The frequency components that are present in the voltage at the inverter bridge, i.e. $\underline{U}_2$ and also in the grid-side current, i.e. $\underline{I}_1$, are not found in the voltage at the PoC, i.e. here called more general $\underline{U}_1$, for an ideal network impedance (test stand or LV network) being zero. As explained in chapter 3, these current components originate from the convolution of the signals at the inverter bridge and possibly from control algorithms. These components are present in $\underline{Y}_{12}$ and can be seen as the harmonic contribution of the device. The distortion in $\underline{U}_1$ originates from both, the distortion caused by the power grid and the distortion caused by the inverter. However, the passive and linear characteristics of the filter circuit implies that the filter circuit can only propagate a distortion but does not create distortion components at new frequencies. For the contribution assessment, the share of $\underline{U}_2(f)\underline{Y}_{12}(f)$ and the share of $\underline{U}_1(f)\underline{Y}_{11}(f)$ with regard to the distortion of $\underline{I}_1(f)$ can be taken under consideration in future studies of single-phase inverters. [8]

D.2 DC-link capacitance

As demonstrated in chapter 3, the DC-link capacitance is one of the main parameters that affect the harmonic current spectrum. Also, together with the previously introduced AC-side filter circuit identification, the capacitance of the DC-link capacitors is an important component for device modeling that affects the voltage ripple of the DC-link voltage. The following method in this section has been disclosed by the author in [9] and presents the remaining final steps with regard to the analytic description in chapter 3 to identify the DC-link capacitance and to validate the method.

D.2.1 Condition monitoring

Typically, the analysis and the surveillance of certain relevant device components are part of the condition monitoring (CM). In theory, there are established methods for CM of the DC-link capacitance. These methods can be categorized into non-invasive and invasive methods based on current sensors, into circuit-based methods and into data-based methods, e.g. according to [94]. The established non-invasive approach measures the currents through the capacitors to identify the equivalent series resistance (ESR) and the capacitance [95, 96]. Further methods have been proposed, e.g. based on a variable electrical network (VEN) [97] to perform the CM during shutdowns. For rectifiers, specific prototypes have been designed and built to identify the DC-link capacitance at 0° and at 90° of the PoW of the voltage at power frequency (50 Hz) [98]. However, both methods are not applicable as they require the specifically desgined prototype devices to perform the method.

D.2.2 Theory

The power resulting from the DC-side PV strings p_{PV}, e.g. as shown in figure 2.2, is set constant for the measurement-based identification is in the laboratory. Based on section 3.4.1,

$$C_{\mathrm{DC}} = \frac{p_{\mathrm{PV}}}{4\pi f_1 u_{\mathrm{DC\,avg}} \hat{u}_{\mathrm{DC\,rip}}} \tag{D.13}$$

can be formulated and describes the DC-link capacitance in general. The main idea of the approach is to make use of the relation between the amplitude of the ripple voltage $\hat{u}_{\mathrm{DC\,rip}}$, the power flow from the DC side of the inverter to the AC side and consequently the time-dependent energy stored in the capacitance and the capacitance itself. For a sinusoidal voltage at the PoC, figure D.6 presents the point measurements at both sides of the capacitor bank (all DC-link capacitors wired together), i.e. $u_{\mathrm{DC}}(t)$ in the simulation. The DFT is based on 5 cycles of the voltage fundamental frequency, i.e. 50 Hz.

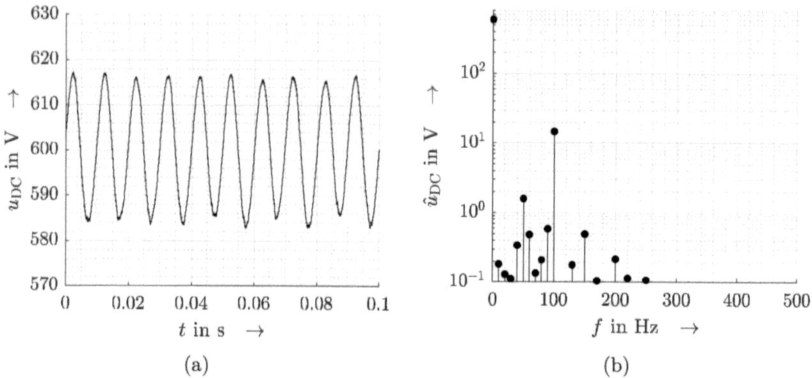

Figure D.6: Time-domain values (a) and amplitude values of discrete fourier transform data up to 500 Hz (b) of voltage u_{DC} across the DC-link capacitance in steady state [9].

D.2.3 Simulative validation

With the described theory and the analytic pre-considerations in chapter 3, a simulation is performed with the same model as already for the AC-side filter circuit identification (see section A.1.1) to validate the proposed method. According to table A.3, the DC-link capacitance is 400 µF. The theory and the simulation neglect the voltage drop over the on-state resistance when the IGBTs are conducting.

In practice, the losses are not zero and the voltage u_{DC} is typically not measureable. The voltage at the inverter bridge u_{Inv} that represents the chopped DC-link voltage u_{DC} is typically available and the losses of modern inverters are neglible. The simulated voltage across the semiconductors, i.e. the voltage u_{Inv} is represented in figure D.7. The similarity of the envelope of u_{Inv} and u_{DC} become obvious when comparing figure D.6 and figure D.7. While a DFT of u_{Inv} does not directly provide the same amplitudes as the DFT of u_{DC} due to the reduced energy resulting from the chopping, the time-domain evaluation of the envelope presents the same amplitude of the dominant ripple component at $2f_1$.

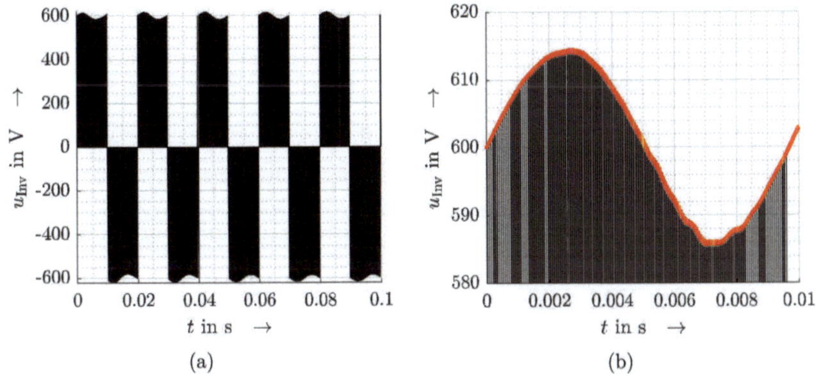

Figure D.7: Time-domain values (a) and detail of time-domain values (black) and respective envelope (red) (b) of simulated voltage u_{Inv} based on [9].

Consequently, $u_{\mathrm{DC\,rip}}(t)$ can be reconstructed based on $u_{\mathrm{Inv}}(t)$.

For this simulation, the dominant $2f_1$ component is identified with 14.91 V. More detailed studies of the performed study have shown that, besides the DC-component, the f_1 component (i.e. the 100 Hz component) is larger than the voltage components at other frequencies by more than a factor of ten.

Based on the voltages and currents at the AC-side inverter terminals, the active power P_{PV} and reactive power Q_{PV} are plotted in figure D.8 The active power averaged over

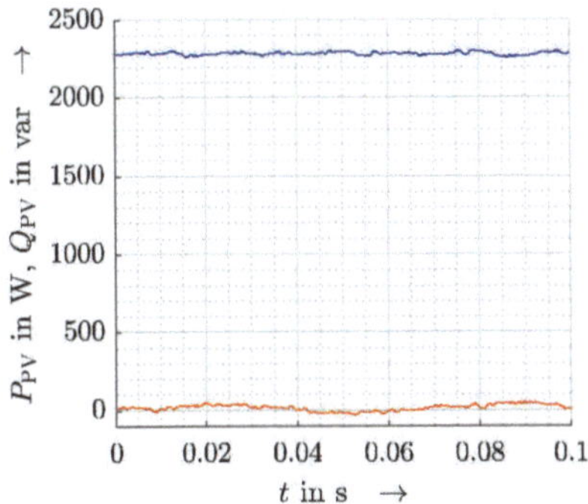

Figure D.8: Active power P_{PV} (blue) and reactive power Q_{PV} (red) at AC-side inverter terminals based on [9].

a 100 ms-window amounts 2280 W. The averaged DC-link voltage $u_{\mathrm{DC\,avg}}$ (also 100 ms-window) is 600.01 V and the power frequency is exactly 50 Hz.

According to (D.13), the DC-link capacitance can calculated based on the proposed method [9] and the respecitve simulation values with

$$C_{\mathrm{DC\,sim\,1}} = \frac{2280\,\mathrm{W}}{4\pi 50\,\mathrm{Hz} \cdot 600.01\,\mathrm{V} \cdot 14.91\,\mathrm{V}} = 405.63\,\mu\mathrm{F}. \tag{D.14}$$

Comparing this value with the implemented 400 µF, a difference of about 1 % results from small deviations in the power, e.g. the presence of a reactive power whereas the theory assumes a reactive power of zero. In total, the deviation of 1 % seems acceptable for the author to consider the presented method based on the theoretic approach as valid.
For the assessment of the impact of the operating power, the results of a second simulation have been evaluated. In this second simulation, the operating power has been reduced by 50 % to 1100 W. The maximum DC-link voltage has been identified for this second simulation with 607.0 V and the minimum DC-link voltage with 592.8 V. The identified capacitance value has been identified with 410.96 µF wich implies a deviation of 2 % of the real value and a deviation of 1 % from the identified value at 100 % of the rated operating power. Consequently, the method should be applied for an inverter operating as close as possible to 100 % of its rated power.
A second model of a single-phase inverter has been studied and is available in the MATLAB/Simulink library. This model is designed to operate at a power frequency of 60 Hz. The DC-link capacitance of this inverter is configured with 3 mF. The result of the respective identification based on the applied method and the simulation is calculated with

$$C_{\mathrm{DC\,sim\,2}} = \frac{2620\,\mathrm{W}}{4\pi 60\,\mathrm{Hz} \cdot 434.25\,\mathrm{V} \cdot 2.75\,\mathrm{V}} = 2.91\,\mathrm{mF}. \tag{D.15}$$

This presents a deviation of about 3 % compared to the implemented capacitance in the model. [9]

D.2.4 Measurements

The laboratory setup is similar to the set up in chapter 3 for the identification of the inverter impedance, i.e. no intended impedance is applied at the AC-side inverter terminals. The measurements are performed with inverter II and have to be performed with an open case similarly to the previously introduced AC-side filter identification as shown in figure D.3. The intended voltage that is controlled by the programmable power amplifier is sinusoidal with an amplitude value of 325 V at 50 Hz. The measurement equipment as described in section B is used with the respective measurement uncertainties. The value-specific measurement uncertainties are below 1 % for $u_{\mathrm{DC\,av}}$ and $\hat{u}_{\mathrm{DC\,rip}}$, so that with a Taylor approximation of 1st degree, p_{PV} is measured with a maximum uncertainty of 2 % and the value of C_{DC} is identified with a maximum uncertainty of 4 %. This uncertainty is much lower than typical uncertainties based on manufacturing tolerances related to the

data sheet values, e.g. around 20 % in table D.1. In addition, the ambient temperature can have an effect on the capacitance but the resulting deviation is assumed below 0.8 % according to [95].
The measurement results are shown in figure D.9. The overvoltages are clearly visible in

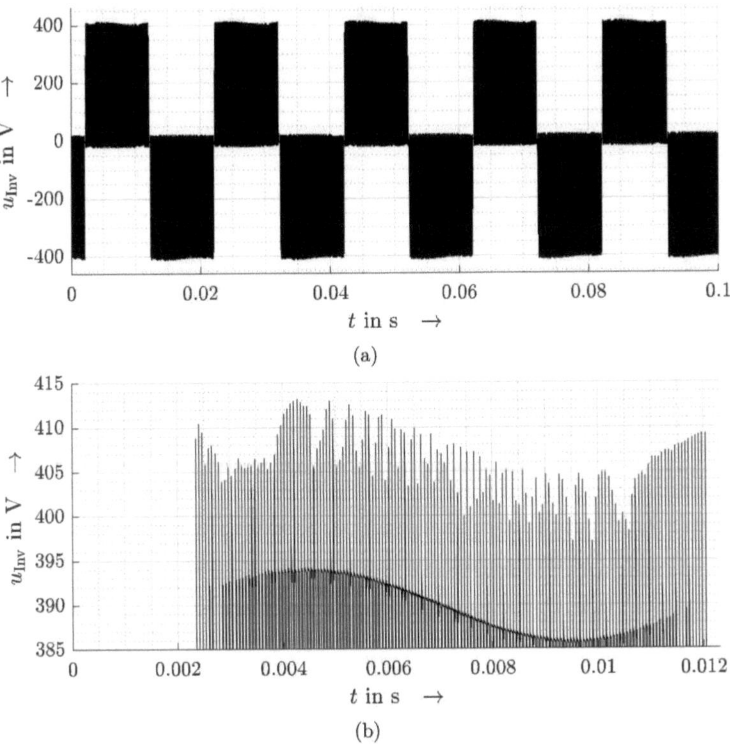

Figure D.9: Measured voltage u_{Inv} (a) and a respective detail (b) of inverter II [9].

figure D.9 (b) and result from the real switching behavior including parasitic effects of the IGBTs that have been neglected partially in the simulation based on ideal switching. The amplitude of the dominating $2f_1$-component that is identified based on the envelope of $u_{\mathrm{DC}}(t)$, e.g. $u_{\mathrm{DC\,rip}}(t)$, is identified with an approximate value of 4 V and the averaged DC-link voltage $u_{\mathrm{DC\,av}}$ with 390 V. The operating power p_{PV} generated by the PV modules is generated by the DC power source with 2300 W. The capacitance of the measured inverter results in

$$C_{\mathrm{DC\,meas}} = \frac{2300\,\mathrm{W}}{4\pi 50\,\mathrm{Hz} \cdot 390\,\mathrm{V} \cdot 4\,\mathrm{V}} = 2.347\,\mathrm{mF} \tag{D.16}$$

based on D.13. [9]
The identified capacitance of the commercially available inverter is larger, i.e. the inverter design is more conservative than the capacitance in the simulation model. Both values are in the typically implemented range for single-phase low power inverters. In general, a large DC-link capacitance is more expensive but reduces the amplitude $\hat{u}_{\mathrm{DC\,rip}}$ of the DC-link voltage and consequently the stress on the DC-link capacitors. A smaller DC-link capacitance requires a faster control but reduces the costs for the capacitors. An appropriate trade-off is left to the device designer.

Method advantages

As required, the presented method [9] is non-invasive and does not require a current injection or a modification of the software, i.e. the control parameters, or the hardware. The method can be applied after completing the manufacturing process. Besides the parameter identification for grey-box models, this enables also to monitor and check the quality of the production process, e.g. the mechanical connection and parasitic and unwanted behavior. Furthermore, aging can be assessed without taking apart the device.
Also, the method is also applicable similarily for PE devices with higher power ratings and high DC-link voltage levels, even during the normal operation without taking the device out of service. No device component needs to be removed thus the risk of a destruction of the device is minimized [9].

Impact of inverter losses

Besides the measurement uncertainties, the losses inside the inverter p_{loss} have been mentioned. To consider the impact of the losses in the calculation of C_{DC}, (D.13) can be reformulated to

$$C_{\mathrm{DC}} = \frac{p_{\mathrm{PV}} - p_{\mathrm{loss}}}{4\pi f_1 u_{\mathrm{DC\,av}} \hat{u}_{\mathrm{DC\,rip}}}. \tag{D.17}$$

Maximum losses inside grid-connected inverters are typically below 6 % [99]. Neglecting these losses will affect the accuracy of the proposed method directly, i.e. by maximum 6 %. [9]

Modelling and stability considerations

The DC-link capacitance is a main component regarding harmonic studies and the harmonic stability. Many white-box simulations neglect the DC-link capacitance and represent the DC-link voltage as a stiff voltage source. Including the DC-link capacitance allows the DC-link voltage to be more realistic, i.e. non-stiff and consequently includes the DC-link voltage ripple. As mentioned previously in this work but also in [100], neglecting the frequency coupling components, e.g. resulting from the DC-link voltage ripple,

by simplifying the model to an LTI model can cause misleading statements towards the harmonic distortion at the PoC and the harmonic stability of the inverter.
In detail, the time variation of the DC-link voltage, also more generally reffered to as DC-link voltage dynamics, can cause a negative real part of the input impedance of an inverter or rather a PE device in general ([75, 101]).
While the mentioned studies in this section are white-box studies and require *a priori* knowledge about the size of the DC-link capacitance, the new, presented method allows to upgrade the current black-box models to grey-box models by using classical electric network elements.

D.3 Control identification considerations

As previously mentioned, there is currently no method available that can identify an unknown topology and the respective parameters of the inverter control non-invasively.
Some ideas to advance on the grey-box modeling are described in the following.
In future, a so-called rejection ratio that studies the active frequency region of the inverter could possibly be used to develop an equivalent model of the control. The fundamental idea is that the parameters of the hardware components do not change. The controller parameters also do not change. However, the impact of the control will be limited to its bandwidth that is defined by the bandwidth of the control. In general, the power dependency results from the share of the control and its respective transfer function and the passive hardware components, e.g. the AC-side filter circuit. If the control reference signals are larger, i.e. for higher powers, the share of the control will be more dominant in the aggregated inverter impedance. This also explains the larger low-frequent active frequency range for higher powers. If suitable fitting algorithms can identify the power-dependent rate of change of the inverter admittance, the control can be represented in the frequency range of its bandwidth, i.e. below switching frequency, as a block, e.g. in a signal flow chart, that reflects the power dependency.

D.4 Advances on grey-box modeling

With the validated identification methods of the AC-side filter circuit and the DC-link capacitance, the control block can possibly also be identified or rather its parameters fitted. The resulting grey-box model would consequently be based on three blocks. The identified AC-side filter circuit model can be transferred into time domain and the capacitance can generally be represented in time domain.
Based on the previous findings of chapter 3, the overall inverter can be represented as a power-dependent LTP model.

If an identified representation of the control algorithm, e.g. by transfer functions, is transferred into time domain, a time-domain grey-box model can be generated thus enabling time-domain simulations of commercially available inverters with an unknown design.
Besides the mentioned identification of the device control, the final step to complete the grey-box modeling approach is the validation of the measurements in cooperation with manufacturers. This would allow to measure the individual component parameters separately, know the control parameters and compare these known values to the parameters that are identified based on the measurements. However, the detaild design parameters of the devices are typically not disclosed by the manufacturers and kept a secret. This validation requires cooperative manufacturers and is left to future work.

Appendix E

Generalized parametrization of linear time periodic systems

As an extension of chapter 4, the following provides the background theory for future non-invasive field measurements.

E.1 Floquet theory

If periodicity is formulated as a function fcn of time t for a period duration T with

$$\mathrm{fcn}(t) = \mathrm{fcn}(t + T), \tag{E.1}$$

according to Floquet theory [102], the second order ordinary differential equation (ODE)

$$\frac{\mathrm{d}^2 y}{\mathrm{d}t^2} + \mathrm{fcn}(t) y = 0, \tag{E.2}$$

also known as Hill differential equation [103], can be represented as

$$y(t + T) = \sum_d a_d y(t) \tag{E.3}$$

and is well-known in form of the Mathieu equation [104]

$$\frac{\mathrm{d}^2 y}{\mathrm{d}t^2} + (\delta + 2\epsilon \cos 2t) y = 0 \tag{E.4}$$

with the parameters δ and ϵ. Introducing the unitary Floquet operator $\hat{F}$ in terms of

$$\hat{F} = \hat{U}(t + T, t) \tag{E.5}$$

and Eigenvalues of one,

$$\hat{F}(t)\Psi_\alpha(t) = \mathrm{e}^{-\frac{\mathrm{j}\epsilon_\alpha T}{\hbar}} \Psi_\alpha(t) \tag{E.6}$$

can be formulated with the Floquet states Ψ_α, the Eigenstates ψ_α and the parameter $\hbar \in \mathbb{Z}^*$ by

$$\Psi_\alpha(t) = \mathrm{e}^{-\frac{\mathrm{j}\epsilon_\alpha t}{\hbar}} \psi_\alpha(t) \tag{E.7}$$

and respectively with

$$\psi_\alpha(t+T) = \mathrm{e}^{-\frac{\mathrm{j}\epsilon_\alpha t}{\hbar}} \mathrm{e}^{-\frac{\mathrm{j}\epsilon_\alpha T}{\hbar}} \psi_\alpha(t+T) \tag{E.8}$$

for real ϵ_α, so that the Floquet states can be factorized into a phase and a time-periodic part with

$$\psi_\alpha(t+T) = \psi_\alpha(t) \tag{E.9}$$

and

$$\hat{H}(t)\mathrm{e}^{-\frac{\mathrm{j}\epsilon_\alpha t}{\hbar}} \psi_\alpha(t) = -\mathrm{j}\hbar \left(-\frac{\mathrm{j}}{\hbar}\epsilon_\alpha \mathrm{e}^{\frac{\mathrm{j}\epsilon_\alpha t}{\hbar}} \psi_\alpha(t) + \mathrm{e}^{\frac{\mathrm{j}\epsilon_\alpha t}{\hbar}} \frac{\mathrm{d}}{\mathrm{d}t} \psi_\alpha(t) \right) \tag{E.10}$$

with the Hamilton operator $\hat{H}$ or rather

$$\left(\hat{H}(t) - \mathrm{j}\hbar \frac{\mathrm{d}}{\mathrm{d}t} \right) \psi_\alpha(t) = \epsilon_\alpha \psi_\alpha(t) \tag{E.11}$$

for a Hermitian operator $\hat{K}$ with

$$\hat{K}(t) = \hat{H}(t) - \mathrm{j}\hbar \frac{\mathrm{d}}{\mathrm{d}t} \tag{E.12}$$

by following the Schrödinger's equation.

Applying the Floquet theory to the field of classical mechanics, $\hat{H}$ can become time-independent for the canonical coordinates p and q, if time is introduced as a new coordinate t' and a canonical impulse $p_{t'}$, with

$$p_{t'} = -\mathrm{j}\hbar \frac{\mathrm{d}}{\mathrm{d}t}, \tag{E.13}$$

such that

$$\hat{K}(p, q, p_{t'}, t') = \hat{H}(p, q, t') + p_t' \tag{E.14}$$

results in

$$\frac{\mathrm{d}q}{\mathrm{d}t} = \frac{\mathrm{d}\hat{K}}{\mathrm{d}p} = \frac{\mathrm{d}\hat{H}}{\mathrm{d}p} \tag{E.15}$$

and

$$\frac{\mathrm{d}p}{\mathrm{d}t} = -\frac{\mathrm{d}\hat{K}}{\mathrm{d}q} = -\frac{\mathrm{d}\hat{H}}{\mathrm{d}q} \tag{E.16}$$

and consequently

$$\frac{\mathrm{d}t'}{\mathrm{d}t} = \frac{\mathrm{d}\hat{K}}{\mathrm{d}p_{t'}} = 1 \tag{E.17}$$

and

$$\frac{\mathrm{d}p_{t'}}{\mathrm{d}t} = -\frac{\mathrm{d}\hat{K}}{\mathrm{d}t'} = -\frac{\mathrm{d}\hat{H}}{\mathrm{d}t'}. \tag{E.18}$$

This finding concludes that explicit time-dependent systems with one degree of freedom can be reformulated into time-invariant systems with two degrees of freedom, e.g. the extended phase space is four dimensional $(p, q, p_{t'}, t')$. More general, the time-dependency of periodic systems is related to the dimension of the phase space.

E.2 Multi-frequent parametrization with superimposed frequency excitation

The previous findings of resolving the time-periodicity of linear systems by increasing the order of the degrees of freedom for the system representation can also be used for the system identification. If the original system is considered by its new representation, the time-constant elements of the new system representation can be identified by using matrix operations with its respective restrictions.

E.2.1 Theory

The main advancement of the following theory is to detach from the vectorial representation of the input and output signals. The implication that each input-output pair has to provide a specific information individually and contribute to the identification of individual elements is replaced by considering the entity of all input vectors in terms of a matrix to identify the entire LTP system. The dimensions of the input and output data are now increased by one. In matrix notation,

$$\boldsymbol{I}_{\mathbf{LTP}} = \boldsymbol{Y}_{\mathbf{LTP}} \boldsymbol{U}_{\mathbf{LTP}} \tag{E.19}$$

can be formulated for the input vectors aggregated in the matrix ${\boldsymbol{U}_{\mathbf{LTP}}}^{NxM}$ and the output vectors aggregated in the matrix ${\boldsymbol{I}_{\mathbf{LTP}}}^{KxM}$. The number of performed measurements is represented by M, the number of applied voltage frequencies by N and the number of the studied current frequencies by K and results in detail in

$$
\begin{aligned}
&\begin{pmatrix}
\underline{Y}_{\mathrm{LTP}\,1\,1} & \underline{Y}_{\mathrm{LTP}\,2\,1} & \cdots & \underline{Y}_{\mathrm{LTP}\,N\,1} \\
\underline{Y}_{\mathrm{LTP}\,1\,2} & \underline{Y}_{\mathrm{LTP}\,2\,2} & \cdots & \underline{Y}_{\mathrm{LTP}\,N\,2} \\
\vdots & \vdots & \ddots & \vdots \\
\underline{Y}_{\mathrm{LTP}\,1\,K} & \underline{Y}_{\mathrm{LTP}\,2\,K} & \cdots & \underline{Y}_{\mathrm{LTP}\,N\,K}
\end{pmatrix}
\begin{pmatrix}
\underline{U}_{\mathrm{LTP}\,1\,1} & \underline{U}_{\mathrm{LTP}\,2\,1} & \cdots & \underline{U}_{\mathrm{LTP}\,M\,1} \\
\underline{U}_{\mathrm{LTP}\,1\,2} & \underline{U}_{\mathrm{LTP}\,2\,2} & \cdots & \underline{U}_{\mathrm{LTP}\,M\,2} \\
\vdots & \vdots & \ddots & \vdots \\
\underline{U}_{\mathrm{LTP}\,1\,N} & \underline{U}_{\mathrm{LTP}\,2\,N} & \cdots & \underline{U}_{\mathrm{LTP}\,M\,N}
\end{pmatrix} \\
&= \begin{pmatrix}
\sum_{n=1}^{N} \underline{I}_{\mathrm{LTP}\,n\,1\,1} & \sum_{n=1}^{N} \underline{I}_{\mathrm{LTP}\,n\,1\,2} & \cdots & \sum_{n=1}^{N} \underline{I}_{\mathrm{LTP}\,n\,1\,M} \\
\sum_{n=1}^{N} \underline{I}_{\mathrm{LTP}\,n\,2\,1} & \sum_{n=1}^{N} \underline{I}_{\mathrm{LTP}\,n\,2\,2} & \cdots & \sum_{n=1}^{N} \underline{I}_{\mathrm{LTP}\,n\,2\,M} \\
\vdots & \vdots & \ddots & \vdots \\
\sum_{n=1}^{N} \underline{I}_{\mathrm{LTP}\,n\,K\,1} & \sum_{n=1}^{N} \underline{I}_{\mathrm{LTP}\,n\,K\,2} & \cdots & \sum_{n=1}^{N} \underline{I}_{\mathrm{LTP}\,n\,K\,M}
\end{pmatrix}.
\end{aligned}
\tag{E.20}
$$

In theory, it is possible to relate the current components at one frequency to the voltage components at different frequencies, though the measurements can only record the resulting currents at each current frequency for each measurement. The current components at each frequency can be aggregated by adding the individual summands to sumcurrents with

$$
\begin{aligned}
&\begin{pmatrix}
\sum_{n=1}^{N} \underline{I}_{\mathrm{LTP}\,n\,1\,1} & \sum_{n=1}^{N} \underline{I}_{\mathrm{LTP}\,n\,1\,2} & \cdots & \sum_{n=1}^{N} \underline{I}_{\mathrm{LTP}\,n\,1\,M} \\
\sum_{n=1}^{N} \underline{I}_{\mathrm{LTP}\,n\,2\,1} & \sum_{n=1}^{N} \underline{I}_{\mathrm{LTP}\,n\,2\,2} & \cdots & \sum_{n=1}^{N} \underline{I}_{\mathrm{LTP}\,n\,2\,M} \\
\vdots & \vdots & \ddots & \vdots \\
\sum_{n=1}^{N} \underline{I}_{\mathrm{LTP}\,n\,K\,1} & \sum_{n=1}^{N} \underline{I}_{\mathrm{LTP}\,n\,K\,2} & \cdots & \sum_{n=1}^{N} \underline{I}_{\mathrm{LTP}\,n\,K\,M}
\end{pmatrix} \\
&= \begin{pmatrix}
\underline{I}_{\mathrm{LTP}\,1\,1} & \underline{I}_{\mathrm{LTP}\,1\,2} & \cdots & \underline{I}_{\mathrm{LTP}\,1\,1} \\
\underline{I}_{\mathrm{LTP}\,1\,2} & \underline{I}_{\mathrm{LTP}\,2\,2} & \cdots & \underline{I}_{\mathrm{LTP}\,2\,M} \\
\vdots & \vdots & \ddots & \vdots \\
\underline{I}_{\mathrm{LTP}\,K\,1} & \underline{I}_{\mathrm{LTP}\,K\,2} & \cdots & \underline{I}_{\mathrm{LTP}\,K\,M}
\end{pmatrix}.
\end{aligned}
\tag{E.21}
$$

To identify the system matrix $\mathbf{Y_{LTP}}$, (E.19) can be reformulated to

$$
\mathbf{Y_{LTP}} = \boldsymbol{I}_{\mathbf{LTP}} \boldsymbol{U}_{\mathbf{LTP}}{}^{-1}. \tag{E.22}
$$

Though (E.22) is straightforward, the calculation of the inverse $\boldsymbol{U}_{\mathbf{LTP}}{}^{-1}$ entails certain restrictions, e.g. the specifications of the input vectors. Calculating the inverse of the input matrix $\boldsymbol{U}_{\mathbf{LTP}}{}^{-1}$ requires the determinant $\det(\boldsymbol{U}_{\mathbf{LTP}})$ to be non-zero and $\boldsymbol{U}_{\mathbf{LTP}}$ being quadratic, so that the rank of $\boldsymbol{U}_{\mathbf{LTP}}$ has to fulfill

$$
\operatorname{rank}(\boldsymbol{U}_{\mathbf{LTP}}) = N = M \tag{E.23}
$$

and thus determines the solvability as well as the dimensions of $\boldsymbol{Y}_{\mathbf{LTP}}$. The inverse can be calculated via the adjunctive $\mathrm{adj}(\boldsymbol{U}_{\mathbf{LTP}})$ in terms of

$$\boldsymbol{U}_{\mathbf{LTP}}{}^{-1} = \frac{1}{\det(\boldsymbol{U}_{\mathbf{LTP}})} \mathrm{adj}(\boldsymbol{U}_{\mathbf{LTP}}). \tag{E.24}$$

The proposed method can be used for all systems that can be linearized as an LTP system, i.e. the principle of superposition is applicable for the linearized operating point such that a deviation of the input vector must not change the system matrix.

E.2.2 General comparison

To show the applicability of the method and to draw a comparison to the state of the art method in terms of the frequency sweep as an elementwise identification method (e.g. see section 2.2.3), the new method is exemplarily applied to a system with $\boldsymbol{Y}_{\mathbf{LTP}}{}^{3x3}$. According to (E.23), the rank of $\boldsymbol{U}$ has to be three and consequently the determinant of $\boldsymbol{U}_{\mathbf{LTP}}{}^{3x3}$ is presented by

$$\begin{aligned}\det(\boldsymbol{U}_{\mathbf{LTP}}) =& \underline{U}_{\mathrm{LTP}\,11}\underline{U}_{\mathrm{LTP}\,22}\underline{U}_{\mathrm{LTP}\,33} + \underline{U}_{\mathrm{LTP}\,21}\underline{U}_{\mathrm{LTP}\,32}\underline{U}_{\mathrm{LTP}\,13} + \underline{U}_{\mathrm{LTP}\,31}\underline{U}_{\mathrm{LTP}\,12}\underline{U}_{\mathrm{LTP}\,23} - \\ & \underline{U}_{\mathrm{LTP}\,31}\underline{U}_{\mathrm{LTP}\,22}\underline{U}_{\mathrm{LTP}\,13} - \underline{U}_{\mathrm{LTP}\,21}\underline{U}_{\mathrm{LTP}\,12}\underline{U}_{\mathrm{LTP}\,33} - \underline{U}_{\mathrm{LTP}\,11}\underline{U}_{\mathrm{LTP}\,32}\underline{U}_{\mathrm{LTP}\,23}.\end{aligned} \tag{E.25}$$

The number of studied frequencies K is also set to three to compare to the frequency sweep by the aggregated output vectors in $\boldsymbol{I}_{\mathbf{LTP}}{}^{3x3}$. Considering (E.25) and (E.27) in (E.24) for (E.22) simplifies with (E.27) to

$$\begin{aligned}\boldsymbol{Y}_{\mathbf{LTP}} = & \begin{pmatrix} \underline{I}_{\mathrm{LTP}\,11} & \underline{I}_{\mathrm{LTP}\,21} & \underline{I}_{\mathrm{LTP}\,31} \\ \underline{I}_{\mathrm{LTP}\,12} & \underline{I}_{\mathrm{LTP}\,22} & \underline{I}_{\mathrm{LTP}\,32} \\ \underline{I}_{\mathrm{LTP}\,13} & \underline{I}_{\mathrm{LTP}\,23} & \underline{I}_{\mathrm{LTP}\,33} \end{pmatrix} \cdot \\ & \frac{1}{\underline{U}_{\mathrm{LTP}\,11}\underline{U}_{\mathrm{LTP}\,22}\underline{U}_{\mathrm{LTP}\,33}} \begin{pmatrix} \underline{U}_{\mathrm{LTP}\,22}\underline{U}_{\mathrm{LTP}\,33} & -\underline{U}_{\mathrm{LTP}\,21}\underline{U}_{\mathrm{LTP}\,33} & -\underline{U}_{\mathrm{LTP}\,31}\underline{U}_{\mathrm{LTP}\,22} \\ 0 & \underline{U}_{\mathrm{LTP}\,11}\underline{U}_{\mathrm{LTP}\,33} & 0 \\ 0 & 0 & \underline{U}_{\mathrm{LTP}\,11}\underline{U}_{\mathrm{LTP}\,22} \end{pmatrix}\end{aligned} \tag{E.26}$$

as a general calculation rule for $\boldsymbol{Y}_{\mathbf{LTP}}{}^{3x3}$.

$$\operatorname{adj}(\boldsymbol{U}_{\mathbf{LTP}}) = \begin{pmatrix} \underline{U}_{\mathrm{LTP}\,22}\underline{U}_{\mathrm{LTP}\,33} - \underline{U}_{\mathrm{LTP}\,32}\underline{U}_{\mathrm{LTP}\,23} & \underline{U}_{\mathrm{LTP}\,31}\underline{U}_{\mathrm{LTP}\,23} - \underline{U}_{\mathrm{LTP}\,21}\underline{U}_{\mathrm{LTP}\,33} & \underline{U}_{\mathrm{LTP}\,21}\underline{U}_{\mathrm{LTP}\,32} - \underline{U}_{\mathrm{LTP}\,31}\underline{U}_{\mathrm{LTP}\,22} \\ \underline{U}_{\mathrm{LTP}\,32}\underline{U}_{\mathrm{LTP}\,13} - \underline{U}_{\mathrm{LTP}\,12}\underline{U}_{\mathrm{LTP}\,33} & \underline{U}_{\mathrm{LTP}\,11}\underline{U}_{\mathrm{LTP}\,33} - \underline{U}_{\mathrm{LTP}\,31}\underline{U}_{\mathrm{LTP}\,13} & \underline{U}_{\mathrm{LTP}\,31}\underline{U}_{\mathrm{LTP}\,12} - \underline{U}_{\mathrm{LTP}\,11}\underline{U}_{\mathrm{LTP}\,32} \\ \underline{U}_{\mathrm{LTP}\,12}\underline{U}_{\mathrm{LTP}\,23} - \underline{U}_{\mathrm{LTP}\,22}\underline{U}_{\mathrm{LTP}\,13} & \underline{U}_{\mathrm{LTP}\,21}\underline{U}_{\mathrm{LTP}\,13} - \underline{U}_{\mathrm{LTP}\,11}\underline{U}_{\mathrm{LTP}\,23} & \underline{U}_{\mathrm{LTP}\,11}\underline{U}_{\mathrm{LTP}\,22} - \underline{U}_{\mathrm{LTP}\,21}\underline{U}_{\mathrm{LTP}\,12} \end{pmatrix} \tag{E.27}$$

The results according to the frequency sweep at sinusoidal reference can be derived from (E.26) by applying the boundary conditions

$$\underline{U}_{\mathrm{LTP}\,12} = \underline{U}_{\mathrm{LTP}\,13} = \underline{U}_{\mathrm{LTP}\,23} = \underline{U}_{\mathrm{LTP}\,32} = 0 \tag{E.28}$$

and

$$\underline{U}_{\mathrm{LTP}\,11} = \underline{U}_{\mathrm{LTP}\,21} = \underline{U}_{\mathrm{LTP}\,31} \tag{E.29}$$

to the input signal vector. This means that in three consecutive measurements, the amplitude of the first order frequency component, e.g. the power frequency, is always equal while in the 2nd measurement an additional 2nd frequency component and in the 3rd measurement the 3rd frequency component is applied. Including (E.28) and (E.29) into (E.26) results in

$$\boldsymbol{Y}_{\mathbf{LTP}} = \begin{pmatrix} \frac{\underline{I}_{\mathrm{LTP}\,11}}{\underline{U}_{\mathrm{LTP}\,11}} & \frac{\underline{I}_{\mathrm{LTP}\,21}-\underline{I}_{\mathrm{LTP}\,11}}{\underline{U}_{\mathrm{LTP}\,22}} & \frac{\underline{I}_{\mathrm{LTP}\,31}-\underline{I}_{\mathrm{LTP}\,11}}{\underline{U}_{\mathrm{LTP}\,33}} \\ \frac{\underline{I}_{\mathrm{LTP}\,12}}{\underline{U}_{\mathrm{LTP}\,11}} & \frac{\underline{I}_{\mathrm{LTP}\,22}-\underline{I}_{\mathrm{LTP}\,12}}{\underline{U}_{\mathrm{LTP}\,22}} & \frac{\underline{I}_{\mathrm{LTP}\,32}-\underline{I}_{\mathrm{LTP}\,12}}{\underline{U}_{\mathrm{LTP}\,33}} \\ \frac{\underline{I}_{\mathrm{LTP}\,13}}{\underline{U}_{\mathrm{LTP}\,11}} & \frac{\underline{I}_{\mathrm{LTP}\,23}-\underline{I}_{\mathrm{LTP}\,13}}{\underline{U}_{\mathrm{LTP}\,22}} & \frac{\underline{I}_{\mathrm{LTP}\,33}-\underline{I}_{\mathrm{LTP}\,13}}{\underline{U}_{\mathrm{LTP}\,33}} \end{pmatrix}. \tag{E.30}$$

Comparing the results of the newly proposed, general method in (E.30) with the results of the frequency sweep in section 2.2.3, the similarities become obvious. In case the linearization is performed for keeping one frequency component in the input vector constant, the column in $\boldsymbol{Y}_{\mathbf{LTP}}$ that relates to this frequency component by its respective harmonic order is not considered. Hence, the newly proposed method and the state of the art frequency sweep lead to identical results.

E.2.3 Non-invasiveness

In many environments, only certain frequencies of the LTP system are excited and of interest, e.g. in electric public power grids 3rd, 5th, 7th and 9th order voltage harmonics are dominant while most other voltage frequencies can be neglected [53, 105]. Consequently, the LTP model only needs to be identified for the excited frequency components to reflect the system response in this specific environment.
The analytic truncation of order M leads to a quadratic system matrix and the state transition matrix is also quadratic [106], but the system matrix that is identified by least mean square (LMS) or the proposed method does not have to be quadratic under the previously mentioned conditions. This is found unconsidered in various applications, e.g. [107, 108], which leads to erroneous additional requirements for the input-output signal relation or rather the resulting $\boldsymbol{Y}_{\mathbf{LTP}}$. There is no mathematic restriction that demands equidistant frequency steps in the frequency sweep, or an equidistant discrete frequency spectrum of neither the input signal vector for the identification, the output signal vector,

nor regarding the elements of $\boldsymbol{Y}_{\mathbf{LTP}}$. This allows to make use of any naturally occurring system excitation to identify the respective elements of $\boldsymbol{Y}_{\mathbf{LTP}}$ in frequency domain.

E.2.4 Operating point dependency

Different applications have different definitions of their operating points since their parameter set that defines one operating point is specific on the device type. For single-phase PV inverters, the studies in chapter 3 have disclosed that the small-signal model of the inverter is solely dependent on the operating power. Making use of non-invasive techniques can demand long measurement intervals until the inverter has been exposed to the intended expected excitations. The operating power might have changed due to a change in the solar irradiation. Therefore, the measurements have to be related to the operating point at each individual measurement point. However, the operating point can be estimated based on the voltage and current at power frequency, e.g. 50 Hz. Consequently, the non-invasive approach can be applied, which enables an identification of PV inverters in the field, e.g. in public LV networks, where voltage spectra with many frequency components are typically present. A final measurement campaign is required specifically to assesss the sensitivity and the uncertainty of frequency coupling components with small amplitudes.